STARTED SMALL AND JUST GOT *Lucky*

-Living with the Curse of a Curious Mind-

An Autobiographical and Historical Account of a Country Boy's 50-Year Journey Through the Cattle Industry

KENNETH ENG, PH.D

Inquiries should be addressed to

The Eng Family Estate
7970 Fredricksburg Road #101-377
San Antonio, Texas 78229
engnm@hotmail.com

First Edition

ISBN 0-615-98389-9 hdbk

All photographs belong to the author or are used with permission.

Cover design by GeorgeAnne Chalmers

Printed in the United States of America
At OneTouchPoint Ginny's in Austin, Texas

Dedication

To Caroline and the stories we could tell
My Love, Muse, Partner and Soulmate

Caroline McDonald Eng passed away Sunday, June 27, 2010 at University Hospital in San Antonio as the result of a drowning accident. Funeral services were conducted at the First Baptist Church, Bryan, Texas by George Lee McDonald and Dr. Baxter Black.

Caroline grew up on a small family ranch near Bryan, Texas where she worked with her father, competed in rodeos and became a rodeo queen. Following many years in the banking industry, we married, and she became Chief Financial Officer of the Eng Ranches land and cattle operations along with their research and consulting business. Our business enterprises allowed us to travel extensively. Caroline was a full-time working partner, playmate and travel companion.

We resided at our home in San Antonio or our ranch in San Luis Obispo County, California. Caroline's great love of animals was surpassed only by her love for family, friends and husband. She will be remembered as someone who was kind to all and lived life to the fullest.

–K–

The Winning Pair

K. S. Eng

They were just a couple of singles
He enjoyed his space and single life
It took him years to discover
They were meant to be man & wife

They traveled different trails
It was never we, always me
It took a long time to realize
A pair they were meant to be

He was short on patience
She had what he lacked
She was sweet and smooth
They say that opposites attract

What one had never learned
The other was willing to teach
They would scratch each other's back
And that itch that was hard to reach

A king and a queen are singles
As a winning hand that's rare
When they joined hand in hand
They became a winning pair

They weren't always dealt aces
And life wasn't always fair
They made do with love and luck
It was enough for a winning pair

TABLE OF CONTENTS

Foreword

It is interesting to read the life story of someone you thought you knew. It was no surprise that in a one-room grade school in Nebraska he relates that his peers contested who could pee up the most boards on the outhouse wall. With that kind of mentality, stereotypes are expected to appear on a regular basis. He was a horseshoer, rodeo hand, has spent a night in jail, worked on a dude ranch, took French as a second language, is a dedicated gambler and has fought a buffalo in hand-to-hand combat.

A history like that is not normally associated with persons of stability. But then look at his accomplishments; educated at Wayne State University, University of Nebraska, Oklahoma State University and Texas A&M. Over his academic career he was a college professor, a scientific researcher and a salaried nutritionist for industry.

But workin' for wages cramped his style. He let his micturating student mentality rise to the surface. Unleashed, it was free to concentrate on the target that had been flashing in his brain since he had discovered the magic of ruminant nutrition . . . and guess what?! The world of feedlots was beginning its Golden Age!

He joined the elite group of feedlot consultants. Gifted with the ability to sell sand in Saudi Arabia, he engaged feedlot managers, and since he knew what he was talking about, they took him in.

With his willingness to put his money where his mouth is, he began feeding cattle himself. He did not discriminate, his diversity included cow-cow operations, growers and feedlots, even Holstein steers! Kenneth conceded that feedlot cowboys might not like Holsteins in the feedlot because, as he said, "It's hard to chase an animal that insists on following you!"

His personal land and livestock business continued to grow: Nebraska, Oklahoma, California, Texas . . . He called to tell his first wife that he was buying a ranch in New Mexico.

She asked, "Why?"

"Well," he explained, "They have more Ph.D.s and a higher illiteracy rate than any other state."

She replied, "That's because they count you in both categories."

His career has continued to flourish. He surrounds himself with capable people. But he has no grand edifice, no cadre of lawyers in his entourage. He is directly in touch with those he trusts with his business. As he says, "Rule #1, return your phone calls."

Like many who have achieved worldly success, he gives God credit and is a generous man. In 2011 he established and funded the Dr. Kenneth and Caroline McDonald Eng Foundation in honor of his late wife, Caroline.

This book also reveals his gift of writing. It is demonstrated in his ability to write a worthwhile column for thirty years in the publication, *Feedstuffs* and his often funny, touching, self-deprecating, and occasionally eloquent poetry, as you will read in this book.

Kenneth is one of those people who gives back more than he takes and makes our world a better place. His contributions have helped all of us who live in this wonderful world of livestock, to keep agriculture at the forefront of progress, producing safe, abundant food, at reasonable prices for our consumers. He is also a savvy marketing advisor, as demonstrated by his "money where his mouth is," examples.

His latest investment is a new land and livestock operation in Mississippi . . . hummm?

Kenneth had invited his mother to a presentation. As he was on a stage, a young man spoke to his mother and expressed how much he admired Dr. Eng, and how he would "give anything to have a job like his."

She thanked him politely and asked, "I appreciate that, but can you tell me what he does?"

I like Kenny Eng, he's a good guy who can dance with queens, pick up the tab and discuss international finance better than most. But he's still the impish grade school kid who would put a woopie cushion on the ambassador's chair.

If Kenneth had been born 150 years ago, he would have given Charles Goodnight a run for his money!

Baxter Black
2014

Preface

For several years friends and clients encouraged me to write a book on the history of the cattle industry since the 1960's including some personal autobiographical material. Writing is relatively easy for me (I've written over 600 articles, papers, poems, etc.) but I resisted until Caroline drowned. After that, I wanted to document the unique role she played in my life and career and our love for each other and the cattle industry.

Like many youngsters growing up on a High Plains farm at the tail end of the Depression, and dust bowl we were poor but so was almost everyone else so few kids knew the difference. As I began high school, my major interests were hunting, horses, sports and girls. From age 12 to 18, my work revolved around alfalfa and prairie hay. It seems we were always mowing hay, baling or stacking hay, hauling hay or feeding hay in the winter. I sometimes earned $25 per day which I thought was good enough for a career. Had it not been for the mid-'50s drought and some athletic scholarships, I would probably have never left the country for college. During the drought I went east to work at dude ranches which was a new world for me. I was bitten by the traveling bug and that plus the curse of a curious mind resulted in almost constant travel for the next 50 years.

Because of numerous scholarships I had a Ph.D. at 24 and no one was more surprised than I. In the spring of 1962, my family (Pat, Laura and Julie) and I headed south in our blue Biscayne Chevy from Oklahoma State University to Texas. Our destination was Texas A&M Gulf Coast Research Station near Angleton and Freeport. I had no idea that it was the beginning of an incredibly exciting 50-plus years journey. I had just turned 25. This was the beginning of a long, loose association with Texas A&M and several other land grant colleges that continues today through the Dr. Kenneth & Caroline McDonald Eng Foundation.

I've been involved in almost all phases of beef cattle production from academics to research and consulting, cattle feeding, yearling and cow-calf operations and ranch trading. I was usually self-employed and over the years had approximately five different careers without changing job titles. My business philosophy is to remain flexible and have an exit strategy.

The developments and changes in the cattle industry during the past 50+ years, 1960s to 2014, have been incredible and in many ways it's as exciting as the trail drive era from 1865-1890. The reason is simple, cattle people make it that way—or at least they did until the beginning of the new millennium. Not only have there been big changes in the cattle business, there have been changes in people and personalities. I'll let history judge whether this is good or bad, but people and society in general have undergone enormous changes.

When the feedlot industry began expanding in the late '60s and '70s, the boom times were reminiscent of trail drives, gold rushes and oil strikes. "Wide open" would be the best description. Acceptable behavior then might be frowned upon today. In addition to working hard, there were more parties, more whiskey drunk and miles traveled at high speeds than today. Furthermore, there were closer relationships, and it was not unusual for the "lion to lay down with the lamb." Nutritionists and cattlemen associated and got along well with drug companies, feed companies, bankers and even packers. I miss the camaraderie. The term "risk management" did not dominate conversations and if you lost a bundle there was the assumption you'd get an opportunity to recoup. The downside of this scenario was that a lot of people "bought the farm" early because of reckless personal habits and imprudent business practices.

I feel incredibly blessed and fortunate to have been a participant in the cattle industry and to play a role in almost all facets for the past 50 years. I am thankful to my family, good friends, clients and especially Caroline for making it all possible. Others may have accomplished as much, but I seriously doubt anyone had more fun, good times and excitement.

I'm a product of this era, and my career was punctuated by sometimes reckless and rowdy behavior interspersed with hard work, good friends, good luck and good timing. I often marvel I'm still around. Perhaps the Lord still has plans for me and I hope I don't disappoint him.

Although I play a major role in the narrative of this book, it really is not about me as I had my fifteen minutes of fame long ago and many times over. Rather, it's about fascinating people in a fascinating industry during fascinating times. It's also a love story about Caroline who was my muse, fulltime partner and soul mate.

Life goes on without Caroline, and I try to stay busy. Strangely enough, the business part of the equation has gone rather well. I'm less concerned about risk and losing money, and this sometimes improves my business decisions.

Time changes everything; it always has and always will. Those of us who are fortunate to participate in this industry may be often flawed but seldom bored, and we have a lot of stories to tell.

Dr. Kenneth Eng
2014

-CHAPTER 1-

A Rough Start but Lady Luck Met Us at the Hospital

It was a nasty day with lots of snow, but Mom and Dad made it to the Norfolk hospital where I got off to a rough start. To this day, I have an intense dislike for snow and winter that perhaps started that day. They drove in from a small farm eight miles northwest of Newman Grove, Nebraska that is approximately fifty miles west of Norfolk. It was March 1, 1937. That's where I first met "Lady Luck" who has been my frequent companion ever since.

The doctor told my parents I wouldn't last the day, and they should baptize and name me immediately. Thus I received my dad's name, Kenneth, with a Jr. added, and I officially became a Lutheran at a very young age. Both mother and I had septicemia, and I also had diphtheria. Nonetheless, thanks to the Lord and Lady Luck we survived—and as they say, "the rest is history."

The twin disasters of the Great Depression and the dust bowl made for lots of poor people, but none of us kids knew the difference because we were all poor. My early years were great, and as an only child of loving parents and the only small boy of a neighborhood of loving families, I wasn't pampered, however, I was "borderline" spoiled and received lots of attention. I loved it. My twenty years of education (elementary through grad school) interspersed with a variety of jobs was the beginning of an exciting and improbable journey.

Our closest neighbor was Dad's brother and his son, Loren, who is five years older than I, and later his sister, Marlys, who is four years younger. During that time, Marlys sometimes lived with my folks since their mother was in a tuberculosis sanatorium. Loren was the only neighbor kid I had to chum around with, and his main interest was cars and tractors. Mine was dogs, horses and cattle—they were my main companions until I started country school. My dog, Pooch, loved and protected me, but he disliked the rest of the world and that got us both in trouble. All this "came to a head" one day when mother wanted me to come into the house and I ignored her which was a bad decision for everyone concerned. She started pulling me into the house, the dog bit her and chased her around the house and she "tore up" her knee. I remember telling Pooch, "I think we're both in a lot of trouble when Dad comes home"—and we were. I got a whipping and Pooch got shot. That was my first lesson that bad behavior can result in bad consequences. I was sad and mad at the world until we got a new dog. However, I still remember Pooch, my first best friend. When I was three or four years old, another event I remember vividly was an afternoon rainstorm when mother was sitting near the window sewing. Lightning struck the house, started a fire, and knocked mother off her chair and unconscious for a short time. I remember the lightning streaking across the floor and seeing my mother lying there and thinking my world had changed. Fortunately, she regained consciousness, we got the fire out and life went on.

My favorite horse was named Nig, and I didn't know why. It seemed everybody with black horses often had one named Nig. In 1942 just after World War II had started, the Army would send people out to pick up any scrap iron we were able to collect. My mother woke me up around 6:00 AM one summer morning and said "Come quick, I want you to see something." Two Negro soldiers came to pick up our scrap iron, and they were the first Negroes I had ever seen. I began to figure out why our black horse was called Nig, but I doubt anyone else gave it a thought.

Another memory of this time was Loren and I taking lunch to our folks in the hayfield. Loren was nine years old, and he was

Nebraska farm with Pooch

driving his dad's '35 Chevy. We got it up to 60 MPH going across the hay field and I thought, "Wow, I'll never go this fast again." I was wrong because around 1950 or 1951 Loren was driving a 1950 Ford and we got it up to 100 MPH one night. Years later, if Caroline was driving in the country and was in a hurry 120 MPH became the new benchmark.

I started grade school at age five in a one-room country school with eight to ten pupils and one teacher. A couple of years later we moved closer to town and moved up to a two-room country school with sixteen or seventeen pupils. In spite of the fact that most of our teachers were very young and just out of high school or perhaps with a couple of years of training at small college, we had excellent teachers. They made classes interesting. One thing for sure, they had the backing of the school board and all the parents. If you got

Kenneth with pet raccoon

into trouble, there was no doubt whose fault it was, it was yours, not the teacher's. I was busy doing what boys do and got pretty good grades except in conduct which I failed miserably as did most of the boys.

Our spare time in school was spent playing tag, softball or baseball, wrestling, trading odds and ends such as pocket knives, marbles, keys, and so on, and snowball fights in the winter. One of the reasons I enjoyed school was that I had a fast buckskin mare named Bess, and when not racing against other horses, I would race cars or whatever else was handy. I rode two miles each way to school and back which wasn't bad except in the winter. I usually rode bareback because my saddle wasn't very good, and the horse helped keep me warm in the winter. Several of us who owned horses would spend weekends riding around the countryside, pastures and creeks. Occasionally a few girls would join us and that was extra fun, especially when riding double. We had a horse barn at school with a wooden boy's outhouse next to it. One of the boy's diversions when we got bored was to see who could pee up the most boards in the outhouse. Talk about cheap entertainment! Obviously, the tall boys had an advantage.

We went to Newman Grove every Saturday night so the men and women could socialize, buy groceries, sell produce, and the kids could do whatever. We would usually go to Norfolk once a month on Saturdays and that often caused a family argument. Dad wanted to take me to the horse sale that was held Saturday afternoons at the Norfolk Livestock Auction, and mother wanted to take me shopping. I hated shopping because we didn't have much money to buy anything but I had to "try on" something in every clothing store. At that time Norfolk had six clothing stores. However, there came a time when I did need more than my share of new clothes and shoes or boots because I started growing like a weed. When I was thirteen, I was 5'10", and if you could believe it, had size 13 shoes. I'm still 5'10" and fortunately my feet quite growing. For a couple of years, a good portion of our family budget went to replace clothing and footwear I had outgrown.

Kenneth with trick horse

Most of my spare time was spent riding horses, playing either baseball or football or hunting and sometimes trapping. My uncle gave me a .22 Remington pump (1912 model), and my dad gave me his old .16 gauge Iver Johnson shotgun when I was ten. My mother was a great cook and we always had plenty of rabbits and pheasants and occasional squirrels—it's hard to have many squirrels without trees. My trapping project didn't amount to much. I bought several traps plus "how-to-trap" books, but about all I could catch were skunks and raccoons that weren't worth much. The more valuable foxes and coyotes proved to be a lot smarter than me. I started hunting when I was ten and have never quit.

–K–

A Small Boy Twice

K. S. Eng

There isn't very much difference
Between grown men and boys
Except the slze of thelr ego
And the price of their toys

As we travel life's winding road
Wouldn't be nice
If we were a grown man just once
And could be a small boy twice

When you were just a boy
The whole world was new
Your thoughts were mostly innocent
And your friends were usually true

Then you entered manhood
And the winding road got rough
Your frlends became competltors
And you grew hard and tough

When you finally reach the top
You were surrounded by mostly men
Who secretly were wishing
That they were kids again

Your mind became more cluttered
Your aching joints are sore
And in your heart you'd trade it all
To be just a boy once more

I began high school at age thirteen, and there were twenty-eight pupils in our freshman class. I didn't ride my horse to high school because the school was in town and didn't have a horse barn, however, you could get a learner's permit to drive a car. Even though I wasn't the greatest student, I truly enjoyed the next four years. Most of the teachers were pretty good—especially considering how aggravating and mischievous the boys were. I loved football and baseball, sort of enjoyed track and later, really enjoyed boxing. Like most boys I also discovered girls. Ninety percent of the boy's interests and conversations centered around sports, hunting, girls, cars and sometimes horses.

Most of the boys in my class worked hard either before or after school and we had some type of summer work. As long as we worked, most of the parents, City Marshall, other adults and teachers were pretty forgiving of what was sometimes semi-wild behavior. When I was about twelve years old dad, got the idea that we should milk cows. From that time until I left for college, we milked about fifteen cows a day by hand. The only thing modern about our barn was a radio. It was damn cold milking on winter mornings. Dad sold the milk cows shortly after I left for college, but my dislike for milk cows lives on. My motto was "selling milk from a cow is stealing from a baby calf."

When I turned thirteen and for the next several years I worked summers for a neighbor who did custom hay baling with an old Case hand tie baler. I tied bales for 2¢ per bale. After dark or when the dew was heavy, we would load bales on a flatbed and stack them for either feeding later or shipping—mostly to Texas. The drought in the early '50s made Texans good hay customers just as 2010 and 2011 did. Between the 2¢ per bale for tying and 50¢ per hour for loading and stacking bales, I sometimes made $20 to $25 a day which I thought was all the money in the world. In fact, I was thinking of making it a career, but technology raised its ugly head. A few years later most people went to Heston Stacks and hauling stacks in on "low boys." Also, farmers started irrigating land that took hay out of production and turned pasture land into corn and soy beans fields. The cold winter milking and the drought in the mid-fifties made me consider college for the first time. Frankly, we were having so much fun in high school I didn't have time to think about much else, however, most kids were excited about getting out and seeing the world.

Between studies, work and sports, we still found time for a lot of simple entertainment. We also had a lot of luck. A Texan named Jerry Carpenter moved in to teach Veterans Ag, and he started a Junior Legion baseball team for us. We ended up with a four-year record of over 100 wins and less than 10 losses. We won several district tournaments and beat much larger teams such as Norfolk, Columbus, and others. Most of our games were on Saturdays and Sundays, but we somehow still had plenty of time to party on the weekends. I'm not sure how or why, but we got in the habit of drinking a lot of cheap beer and wine which we thought added to the excitement. The beer of choice was Schlitz or Storz and the wine was Mogen David. Their prices fit our pocketbook. Our favorite hangout was a high hill west of Newman Grove where we could drink beer, park with the girls. We called it Poker Hill only no one played cards there. The later I stayed out at night, the earlier Dad would get me up the next morning. One night I didn't get in until 3:00 AM and Dad got me up at 4:00 AM. He told me later that afternoon that we needed to quit this or it was going to kill both of us.

I often marvel at the fact that during my life, the most troublesome and negative events ended up creating a great opportunity. For instance, without the drought in the mid-'50s, I might not have left the country and attended college.

The drought brought grasshoppers and rainmakers and I not certain which were the biggest pest. The rainmakers were going to seed the clouds either by cannon or by airplane although planes were few and far between. Their pitch was that they needed every farmer and merchant in the area to participate so they could get widespread coverage and they charged per acre for their efforts. When they came to our place, Dad declined to participate. They commented that he just wanted all his neighbors to pay but he would still get the benefits. He said, "No that wouldn't be fair, if they are good enough to make it rain, they are smart enough to cut it off at my fence lines, and I'll show them where they are." Dad had a sly sense of humor—often at my mother's expense. One of his favorite stories was that "women did knitting and needlepoint so they would have something to think about while they were talking."

Shortly before high school graduation some of us qualified for the State track meet at the Nebraska University in Lincoln. We did pretty well. I placed in the 220-yard dash and we actually held the state mile relay record for about ten minutes. In the final heat another town broke our record and we ended placing fourth. The highlight of the event, however, was that the University of Nebraska and Kansas were competing in a college meet at the same time. Many are unaware that Wilt Chamberlain also competed in track in college and at that meet, he threw the shot put and ran the 440. I don't recall if he won the 440-yard dash, but watching a seven-foot giant run the 440 was a sight to behold.

Shortly thereafter several of us went back to Lincoln for an Ag Judging contest that ended in a semi-disaster. We had to take two cars and I along with three other fellows stopped in Humphrey, Nebraska and bought two quarts of cheap whiskey. The four of us were staying in the same motel room. On the last night, we decided to party. A group of students from another town asked for a drink and we refused them. A fight ensued and long story short, we pretty much wrecked the motel room. Rather than confessing and taking our lumps before we left, we headed back to Newman Grove early the next morning. Of course the motel owners told our Ag teacher, J. Paul McIntosh. He paid the damages and when he got back to Newman Grove he gathered all of us and read us the riot act. He said he should turn us in to the School Board and our parents, but if we came up with the money for damages he would not. Had he not remained silent, we probably wouldn't have graduated. J. Paul was an extremely smart and a good teacher. He has become a very successful entrepreneur and philanthropist living in

Norfolk, Nebraska. For his sixtieth wedding anniversary, I bought their dinner at Prengers restaurant in Norfolk. J. Paul said, "Kenneth, you didn't have to do that." I said, "Yes, I do because I still owe you one."

J. Paul was just named Nebraska Man of the Year in 2013, and I am proud of him. I called him as soon as I heard and told him I always liked to associate with smart successful people. It is a well-deserved award. I mentioned our Lincoln, Nebraska motel fight and fiasco and he said, "Yeah, I remember and I also remember the first time the Ag Class castrated sheep." We would cut off the bottom of the scrotum, bite and pull the testicles out with our teeth. I don't know where J. Paul found the sheep because none of the students had ever raised any. He reminded me that I went first on the "biting" castration procedure and then others followed. What I remember about that incident was that word got back to the girls in our class and none of us fellows got kissed for at least a week.

Although it didn't seem so at the time, the severe drought and leaving for college was a blessing. It ended an intense relationship with a girl who I really liked even though we didn't have much in common except hormones. We both dodged a bullet. From a boy's standpoint, the decade of the '50s was a great time to grow up. I'm not sure what the girls thought because we were sometimes crude but we sure enjoyed their company.

I stayed on the farm the summer after high school graduation and came back again the next summer. However, with the severe drought there was no way to make a living and after that, so I subsequently spent the next three summers working at dude ranches back east. That was another example of bad luck (the drought) turning into good luck because it exposed me to a world I had never known.

Having survived a rough start and high school, I subsequently hit the ground running and I've been told I've been in a hurry every since.

Engs Have Son — Mr. and Mrs. Kenneth Eng, Newman Grove, are the parents of a son, weighing 6 pounds and 2½ ounces, born Monday morning at 8:15 o'clock in a Norfolk hospital. The condition of the mother and baby was described as "only fair" by attendants Monday noon.

—Mrs. Kenneth Eng and baby were dismissed from a Norfolk hospital Sunday and came home the same day. March 14, 1937

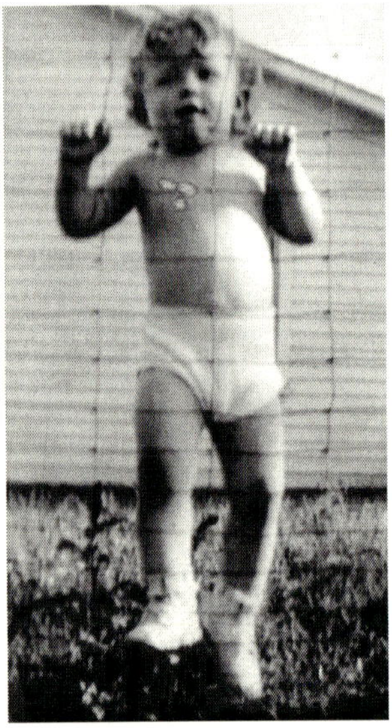

Age 1½, 1938

Age 1, 1938, with Dad

Age 2½, 1938

Age 4, 1941,

Age 12, 1949
8th grade

Age 15, 1953
Junior in high school

Age 15, 1953
High school junior

Age 16, 1954

Age 16, 1949
High school senior

Age 15, 1953

Saturday Night Memories

K. S. Eng

I have so many memories that
will never die
Like working in hay fields from first
to last light
The heat of the sun in the blue summer sky
And heading for town on Saturday night

Daylight to dark we baled and hauled hay
Mom brought lunch and cool lemonade
We would eat lunch and rest at mid-day
With the tractors shadow our only shade

Never thought twice of the hours we'd keep
Loading bales on flatbeds till day's last light
Then crawling into bed and falling asleep
Dreaming of girls and town and
Saturday night

Saturday night we would gather together
And sometimes drink clandestine
homebrew
Talk of girls and sports, and maybe
the weather
And plan our escape to somewhere
exotic and new

We would envy the banker's fancy new car
And talk of what we would someday own
I dreamed of a plane that would
take me afar
But doubted I'd get the money or loan

The best of old friends will someday
be gone
That's the Lord's plan, we've nothing to say
But the best of old memories will linger on
Like boys in the field and the sweet
smell of hay

As my plane flies across the clear
summer sky
And the hay fields slowly come into sight
Memories of past summers refuse to die
As do dreams of town on a Saturday night

We Need You Roy and Gene

K. S. Eng

Remember those Saturday movies with
Roy & Gene
When they battled the bad guys & kept
the town clean
Fair play & justice is what they stood for
They were friends of the rancher,
widows & poor

Politicians & lawyer trembled at their sight
If Roy & Gene caught them there would
be a fight
The shysters & crooks wore mustaches and
black hats
You could tell the good guys from the
crooks & the rats

The bad guys would kick & gouge,
they didn't care
But Roy & Gene would win & always fight fair
Then take care of the widow and hug the
girl in the gown
Kiss Trigger & Champion and ride out of town

That was Saturday night so many years ago
When we cheered our heroes at the
picture show
Fast-forward 50 years to century 21
And you start to wonder is the
West really won

The lawyers and the politicians are
still around
Under each rock a bureaucrat can be found
They are slicker than ever & quick with a pen
They want our freedom and our
land thrown in

They don't use a gun, but with the
pen they're fast
Don't sell them short, look at laws they
have passed
They want control of your water, air
& your dirt
They want your soul now that they have
your shirt

We can win this battle if we make them
fight clean
And if we stand for what's right just like
Roy & Gene
Watch their old movies, get them down
to perfection
And if you kiss your horse, turn him the
right direction

-CHAPTER 2-

I Guess I'll Try College

The summer of 1954 I was offered a scholarship to play football at Wayne State College which was a state college about eighty miles away. I said I would enroll, but I would prefer to box rather than to play football. At that time they had an outstanding boxing team and coach. The football coach said I could do both, but he was lying. I think they had already decided to disband the boxing team because they had a few boxers that after a lot of fights had apparently suffered a little brain damage. That was not a good advertisement for the college. It has always interested me how some fighters can absorb a lot of punishment and never show any ill effects while others can encounter the same or less punishment and suffer minimal or obvious brain damage. At any rate, we boxed quite a bit during the noon hours in high school and I started boxing in the Golden Gloves in 1954 when it was a big deal in our area. We also fought at local carnivals which I really enjoyed. But to make a long story a little longer, I wasn't a very good college football player and suffered a back injury that permanently slowed me down. As a result, I had more time to concentrate on studies and found to my surprise I enjoyed it.

My best piece of serendipitous good luck was meeting a local Wayne girl who was going to be a sophomore. She said, "Kenneth why don't you sign up for Chemistry, and we can be lab partners." I said, "Why not," and even though we didn't have Chemistry at my high school and I didn't know a Bunsen burner from a Erlenmeyer

Referee raising Kenneth's hand at his first Golden Glove bout with Trainer Walt Fick, Inman, Nebraska

flask, I signed up for Chemistry and several other science courses. Another piece of good luck was that the football coach got the football players signed up for what we referred to as the "bonehead" English classes that were usually taught by speech teachers. Thus, my English classes basically were composed of speeches and story writing that helped me later in life. I wasn't a great student and I don't think I'm especially smart, but if I found something interesting I would do well in the class.

This was an era when there were always jobs available for any kids who wanted to work. At Wayne State, I had a job on campus with the State where they paid 75¢ an hour. Then at night I helped out on a farm north of town, and they paid a $1 per hour plus a good evening meal. Sometimes I'd work in the fields all night and get back to the dorm just in time to shower and make it to my first class. The result was that I had all the money I needed for living expenses if I didn't party too much. Working on campus was fun because you usually had either a classmate or teammate as a partner. One winter day while sanding icy streets, my co-worker driving the pickup decided we should park on the hill by the men's dorm and go inside to warm up. Unfortunately, he didn't set the brakes on the pickup and it rolled down hill through Professor Fred Dale's yard where it stopped just in front of his house. I was taking an 8:00 AM Meteorology class from Fred Dale, and the next morning the first thing in class he said, "I'll pay $50 to anyone who will identify who ruined my yard." It really didn't do any damage, but he didn't like jocks and assumed it was someone on the football or basketball team. I could have used the $50 and thought about turning myself in except I needed the class credit.

In another work-related incident, a teammate and I were moving furniture into the basement of a women's dorm. A girl who sat next to me in English class was hanging her clothes on the wash line nearby. This included some panties and bras and my teammate said, "I sure wouldn't mind getting into those." I said, "I'll bet you a case of beer I can get you a pair." We made the bet. I knew Barb, who had beautiful bedroom blue eyes, was almost blind without her glasses. I went over to see her, visited with her for a bit, and I took a pair of her panties and bra and collected my bet. The story had an embarrassing ending because I went home that weekend and my mother decided to clean my car and she found the unmentionables in my glove compartment. Of course she had dad talk to me. He told me, "If you expect your mother to clean your car, check you damn glove compartment first."

During the summer of 1956 it was still so dry Dad didn't need my help on the farm. Even the alfalfa and prairie hay wasn't growing so I couldn't get my usual work in the hay fields. I decided to get out of Nebraska and try the Eastern dude ranch circuit.

I loved my first two years at Wayne State and doubt if I would have made it anywhere else. I call it my personal "head start program." Starting my third year in college I transferred to the University of Nebraska and enrolled in what was called Technical Science Agriculture. It was basically an even split between science classes and general Ag classes. At first I thought I might major in Plant Pathology or Horticulture and I worked in the Horticulture Department and Forestry for two years. Even though they were nice enough, I found those professors a little strange. I later decided that if it didn't grow "hide or hair," I wasn't very interested.

During my time at the University two or three of us fellows always roomed together in a duplex about ten miles from campus. Each of us ended up with Ph.D.'s, and since we all worked in labs, we had access to absolute alcohol (199 proof). Without telling anyone, we bought some Smirnoff Vodka decanters, consumed the contents and kept refilling them with absolute lab alcohol. The result was lots of good parties, friendly girls, and it gave us a definite advantage in poker games.

Perhaps the most outlandish event that occurred at our duplex was what I refer to as our "Great Turkey Hunt." One Saturday night after socializing we were driving near the Ag College and found several turkeys that had escaped from the turkey farm. We confiscated eight of them—by the way, this was no easy task. After wringing their necks and "gutting them" we took them back to the duplex where no one wanted to pick the feathers and dress them. We had an automatic washer and dryer in a laundry room between the two duplex units and I decided to throw the turkeys in the automatic washer. It worked better than you might think. Then we ran the bathtub half full of water and threw the dressed turkeys in the bathtub. However, we didn't give any thought to the residue left in the washing machine. The next morning there was a pounding at the door and I woke up thinking the campus police had discovered the theft and tracked us down. However, when I opened the door one of girls who lived in the other part of the duplex held up her clothes full of turkey feathers and said, "Kenneth, what the hell have you guys done now." I invited her into the apartment and showed her the turkeys in the bathtub and we consequently made a trade. I helped her clean out the washing machine and the girls finished cleaning and dressing the remaining half of the birds.

–K–

The List

K. S. Eng

When I first went off to school
Mom would always say
Mind your manners and study hard
And you'll make the list someday

The list could be one of many
Honor roll or most likely to
She wanted to see my name in print
Any old list would do

When I learned about cars and girls
Academic interests quickly passed
Loved all kinds of wheels and women
Just as long as they were fast

But school is gone and I'm alone
I think of what I missed
Never learned no social skills
Never made no list

My jeans are torn and faded
Wear'm low with no underwear
Won't make that best dressed list
But only mom will care

Can't blame the wife for leaving
And girlfriends are hard to get
I'm doomed to be a bachelor
Never made the most eligible list

Mostly now I go hunting
Me and my one-eyed dog
Last week I bagged a real trophy
Didn't know it was a neighbor's hog

The sheriff, the warden and neighbors
The judge and the highway patrol
Hauled me down to the court house
Cuffed to keep me under control

They said stay out in the country
Don't come to town at night
Stay away from roads and fence lines
Cause there's orders to shoot on sight

Well, mom I hope you're happy
By now you probably guessed
I finally hit the big time
I made the "endangered species" list.

-CHAPTER 3-

Dude Ranch Days
(Go East Young Man)

In 1956, '57 and '58 I spent the summers working at dude ranches, the first year in Florida and then in the Catskills. I would leave Nebraska in mid-May and usually come back in September. Both departure and return dates were somewhat dependent on University class schedules. It was a great experience for a young fellow from Nebraska. Not only were there plenty of horses and girls, but also trees and water that were in short supply where I came from. The Central Florida outfit catered to kids from the Central and South Florida area. They had a fairly decent string of horses and, along with two other wranglers, I took care of them and the riding classes. The ranch also had a lot of other recreation for the kids including swimming, boating, sailing, and fishing. Female counselors were in charge of most of these activities. On days off we would head for one of the beaches (either the Gulf of Mexico or Atlantic Ocean side) and tour the ranch and citrus country along the way. I had no idea there was so much open country and so many cattle in Florida and I truly loved the state. It was also the first time I had seen the Gulf or an ocean.

The highlight of that summer was that I developed a semi-serious relationship with two female counselors, one named Piney and the other Dixie. In addition I had a white mare I was fond of named Tilley. In the heat of battle, so to speak, I would sometimes call one of the girls by the other's name and vice versa—and also slip in the name Tilley. Needless to say my popularity with the two ladies declined and we drifted apart. I wrote a poem entitled "Two Girls and a Mare" which was one of Caroline's favorites. Even though she had a jealous streak, the thought of me getting dumped by two women and a mare appealed to her.

One notable event that occurred on the trip back from Florida is that I was riding with a good friend from Columbus, Mississippi and we were going through Dothan, Alabama late at night. A minor league ballgame had just been finished and the police were directing traffic onto the highway. We noticed we had a tire going flat, and when the traffic cop waved us to stop, we went on assuming he was busy. Wrong! We were at a nearby gas station getting the tire repaired when the policeman drove up with sirens blowing and gun drawn. He screamed he was going to charge us with attempted motor vehicle homicide. He took us down to the jail that was an unpleasant place. In the meantime, we had hidden what little money we had and subsequently, when he went through our billfolds, he said we might rot in jail since we didn't have money for bail. My Mississippi friend who had a great gift of gab (he later became a lawyer) told him that suited us just fine because we were afraid we were going to be drafted into the army, and since it was full of Yankees and had just integrated, that was high on our "not to do list." The cop began talking about his war experiences, and he finally took us back to our pickup and wished us luck. His parting words were "I hope you don't have to go into that damn integrated, Yankee army" (these were not his exact words).

A footnote to the Dothan jail event takes place about twenty years later when I met a fellow in Phoenix who was with the Arizona-Colorado Land and Cattle Company and who had been an officer in the Calvary during World War II. Fred and his wife, Carol, were great people and I believe he was State Senator for quite a while. They later started one of the first river rafting companies going down the Grand Canyon. Fred and I got to talking about our younger years and he related an incident in the late '30s when he was hitchhiking and got thrown in jail in Dothan, and charged with vagrancy. The only difference in his story and mine was that they kept him in jail for 30 days, and he thought it was because they got paid per prisoner for meals. Small world.

My next two summers were spent at a dude ranch in the Catskills about 100 miles north of New York City. This was the highlight of my life up to that point. I would get on a Greyhound bus in Nebraska and go straight through to Grand Central Station, New York City in the spring and return in the fall. The round-trip ticket was $72. Most of the ranch guests were Jewish people from New York City, and this was a new experience for me. I loved the country and our clientele. I concluded that, in general, the people from New York City were nice except that they just didn't like each other. I still rate the Catskills as the prettiest place in the United States—if not the world. I would say the same about some of the young ladies. On that subject, I think we had one Jewish family in Norfolk, Nebraska, and they owned a clothing store. According to my mother and the conventional wisdom of most folks in that area, you could identify Jews because they had larger noses, dark complexion and owned clothing stores. I quickly discovered my mother's description was incorrect, but never told her. Good Lutheran ladies with young sons already had enough to worry about.

The ranch had seven to eight wranglers from all parts of the country and a string of 150 to 175 horses. Each wrangler had approximately twenty-five horses in his string. The horses originally came from the central and western part of the United States via a horse trader who was associated with the original televised rodeo in Cowtown, New Jersey. Because the winters were really severe in that area, all the horses were sold each fall, then we started anew again the following spring season with new horses. Putting together 150 suitable horses from unknown origins for dudes was a challenge.

They were brought up to the ranch in truckloads of approximately thirty head each. We could send back approximately 15%

if we found them unsuitable. Selecting the horses off the truck was a rodeo in itself, and it was great entertainment for the guests. The routine was to unload the horses into the corral, and then we would select horses one at a time. Our choices were based on putting a bridle and saddle on each horse and riding them around the corral once or twice. Hopefully you would get three or four good horses off the load, and you would paint your brand on them. However, a 30-second ride around the corral doesn't guarantee much when it comes to horse flesh. I learned in a hurry to select horses without white hooves or stockings because the country was rocky and rough. Once you put shoes on a horse, you didn't want to spend time re-shoeing because of bad hooves. In addition, a lot of the horses had never been shod before so horseshoeing often became another rodeo. I got pretty good at horseshoeing (this was the strong back/weak mind syndrome) and found a major key to horseshoeing success was to get along with the horse. I became a bit of a "horse whisperer" while other cowboys relied on brute strength. I ended up shoeing quite a few horses for other cowboys in exchange for favors such as liquor, hand tooled belts, and the like.

I concluded the big difference in who could and who couldn't shoe horses was mental. One of the more embarrassing horseshoeing events I was involved in was when the New Jersey horse supplier drove up in his Cadillac along with an Italian-looking driver wearing a continental suit. They had followed a new load of horses. He commented that we'd been sending a lot of horses back and wondered why. Then he pointed to a bay horse that we'd identified to go back on the truck. He said, "What's wrong with him?" I explained, "He's got white hooves and you can't keep shoes on him. and furthermore, he's almost impossible to shoe in the first place."

He motioned to his driver and told him, "Show these country boys how it's done." The fellow took off his suit coat and put on a shoeing apron. He put a lariat around the horse's neck, and he told me to hold the other end. He proceeded to tack on all four shoes in a matter of minutes while the horse never moved a muscle. I sent the horse back anyway. but as Marty Robbins said in one of his songs, "Don't underestimate a cowboy in a continental suit." At that time the going rate for horseshoeing was $4 per hoof plus the cost of shoes. If they brought horses to me, I could easily shoe ten horses per day and at one time thought of making it a career.

We would put on rodeos at the ranch some weekends. These were popular events in the area. One weekend they had a rodeo at the old polo grounds in New York City and some of us were contestants. The "rough stock" events turned out to be a disaster as most of the horses were wild horses from the west that were nasty in the chute and after the chute gate was opened they might just run away. I remember the bareback horse I drew turned around in the chute and bit my left boot really hard. When the chute gate was opened he ran straight away to the center field wall. The only thing I could think of is "say, hey"—this is where Willie Mays once roamed and it was hallowed ground.

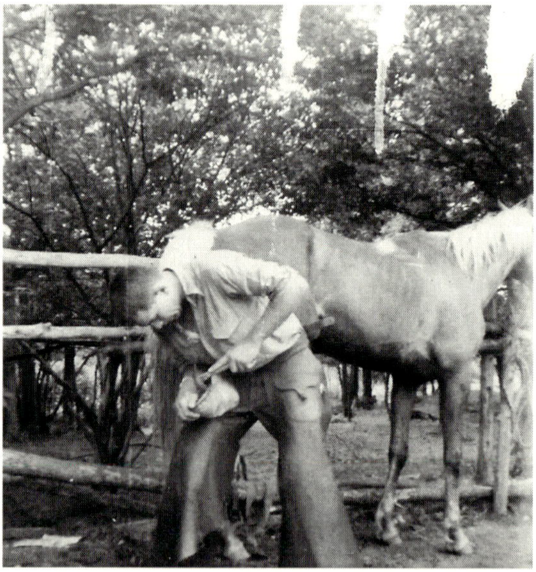

A hard way to make an easy living

Not many points for style, but good enough to win in New York

Riding the Catskill Range

We didn't get a chance to go into New York City much but one morning the boss said if we finished up early he would take us into Manhattan for dinner and a night out. We ate at Jack Dempsey's restaurant. Since I was a boxing fan, it was a little piece of heaven. I kept a photo and the menu as a souvenir. At age 20, I'm the really young, innocent-looking fellow. Others in the photograph are Del Reynolds from Oklahoma, Wes Jackson from Mississippi and Jack Franks, the Dude Ranch owner. The food was great—and take a look at the prices on the 1957 menu!

Dinner at Jack Dempsey's restaurant
Wes Jackson, McComb MS, Jack Franks, dude ranch owner,
K.S. Eng, D. Reynolds, Del City, OK

Jack Dempsey Restaurant menu, 1957

Appetizers

BLUE POINT OYSTER COCKTAIL .70
CHERRYSTONE CLAM COCKTAIL .70
LUMP CRABMEAT COCKTAIL 1 25
JUMBO SHRIMP COCKTAIL 1 10
TOMATO JUICE .35
FRESH FRUIT CUP .50
MARINATED HERRING IN OUR SPECIAL CREAM SAUCE .65
SPECIAL CHOPPED CHICKEN LIVERS .60
BOWL OF CELERY AND MIXED OLIVES .75
CHILLED MELON IN SEASON .55
ICE COLD HONEY DEW MELON OR CANTALOUPE .55

Soups

HOT CONSOMME .40
CREAM OF TOMATO SOUP .60
ONION SOUP AU GRATIN .60
SOUP DU JOUR CUP 35 TERRINE 50
COLD VICHYSSOISE .50 COLD CONSOMME MADRILENE .50

Sea Food

WHOLE BROILED MAINE LOBSTER, Drawn Butter, French Fried Potatoes 3 75

FRESH SHRIMP, PANNED IN BUTTER, NORFOLK STYLE 2 50
French Fried Potatoes, Cole Slaw

FRIED DEEP SEA SCALLOPS 2 10
Bacon, French Fried Potatoes and Cole Slaw

FRESH FILET OF LEMON SOLE, SAUTE MEUNIERE 1 75
Cole Slaw, Parsley Potato

BROILED FRESH GULF STREAM SWORDFISH STEAK 1 95
Maitre d'Hotel, French Fried Potatoes, Cole Slaw

FRIED FRESH FLORIDA SHRIMPS 2 35
Tartar Sauce, Cole Slaw, French Fried Potatoes

MARYLAND SPECIAL LUMP CRABMEAT 2 25
au Gratin in Border of Potato Duchesse

TWO SOFT SHELL CRABS on Toast, Tartar Sauce, Cole Slaw, and French Fried Potatoes 2 50

Vegetables and Potatoes

CORN ON COB .35
BRUSSELS SPROUTS .50 CREAMED SPINACH .50
GARDEN FRESH PEAS .50 STEWED TOMATOES .50
FRESH STRING BEANS .50 FRESH BROCCOLI .50

FRENCH FRIED IDAHO POTATOES .35
HASHED BROWN .40 HOME FRIED .40
HASHED IN CREAM .45 AU GRATIN .45

If for any reason you do not receive the maximum in service and courtesy from my employees, I would appreciate it if you would let me know.
Your host,
Jack Dempsey

Sunday, June 30, 1957

Steaks, Chops, Etc.

As Advertised
BROILED PRIME SIRLOIN STEAK (1 lb.) with Baked Idaho Potato and Our Special Tossed Salad 4 95

BROILED TENDERLOIN STEAK (For One) 4 95
~~BROILED FRESH CALF'S LIVER STEAK~~ 3 10
BROILED SUGAR CURED HAM STEAK 2 75
BROILED PRIME HAMBURGER STEAK with Onions 2 10
BREADED VEAL CUTLET with Tomato Sauce 2 45
BROILED FRESH CORN FED PORK CHOPS 2 25
BROILED FRESH KILLED HALF CHICKEN on Toast 2 15
TWO BROILED DOUBLE SPRING LAMB CHOPS 3 65

(Above Orders Served with French Fried Idaho Potatoes)
(Chef's Salad with Above Orders 45c.)

Salads

EGG SALAD on Crisp Lettuce, Imported Anchovy Fillets, Russian Dressing, Potato Salad 1 75
DEMPSEY'S SALAD BOWL OF MIXED GREENS, Julienne of Assorted Meats with White Meat Turkey 2 55
FRESH CRABMEAT SALAD 2 35 FRESH FRUIT SALAD 1 25
SHRIMP SALAD 2 35 CHICKEN SALAD 1 95
CHEF'S SALAD WITH CHOPPED EGGS .95
IMPORTED HOLLAND HAM, Lettuce and Tomato, Potato Salad, Cole Slaw 1 75
ASSORTED COLD CUTS, Potato Salad, Lettuce, Tomato, Cole Slaw 2 50

SALAD DRESSING:
Mayonnaise 20 Thousand Island 35 Roquefort 45 Russian 25

Sandwiches

TRY OUR DELICIOUS HOT CORNED BEEF or PASTRAMI SANDWICH with Raw Sauerkraut and Dill Pickle .85
PRIME STEAK SANDWICH, GARNI 2 95
DEMPSEY SPECIAL (Turkey, Ham, Swiss Cheese, Sliced Egg and Russian Dressing) 1 65
HAMBURGER with Onions 1 35 HAM AND SWISS CHEESE .95
HAM .75 SWISS CHEESE 75 CLUB 1 45 HOT TURKEY 1 75
CHICKEN SALAD SANDWICH (Dempsey Quality), Mayonnaise 1 10
TUNAFISH (Chicken of the Sea) SALAD SANDWICH .95
LAKE STURGEON, Tomato, Sliced Onion 1 65
COLD SLICED TURKEY SANDWICH, Lettuce and Tomato 1 25

Desserts

GREEN CREAM DE MENTHE (Pony) 55

(Our Delicious Pastries Fresh Daily From Our Own Bake Shop)
LUSCIOUS STRAWBERRY SHORTCAKE .65
FROZEN ICE CREAM CAKE, CHOCOLATE SAUCE .65
FRUIT JELL-O, WHIPPED CREAM OR CHOCOLATE PUDDING .35
FRENCH ICE CREAM OR SHERBET .50
PARFAIT DEMPSEY'S .55 JACK DEMPSEY CHEESE CAKE .50
FRENCH APPLE CRUMB CAKE .40 DANISH PASTRY .25
FROZEN CHOCOLATE ECLAIR .65 CHOCOLATE LAYER CAKE .40
TOASTED HOME MADE POUND CAKE .30 RICE PUDDING .35
CHERRY CHEESE PIE .55 STEWED JUMBO CALIFORNIA PRUNES .40
SLICED FRESH PINEAPPLE in Syrup 40

Cheese

IMPORTED ROQUEFORT .40
CREAM .30 AMERICAN .30
CAMEMBERT OR SWISS .40 LIEDERKRANZ .40
JACK DEMPSEY'S OLD CANADIAN CHEDDAR CHEESE .50

Beverages

CUP OF COFFEE .20 MILK .20 TEA .20
DEMI-TASSE .20 HOT CHOCOLATE .25
SANKA OR POSTUM (Pot) .25 BUTTERMILK .20
FRESH ORANGEADE .45 FRESH LEMONADE .45

Wine and Liquor List

COCKTAILS

Alexander with Gin .75	Clover Club .75	Manhattan .65	Rob Roy .85
Alexander with Brandy .90	Champagne .85	Old Fashioned (Rye) .70	Dempsey's Special .85
Bronx .65	Frozen Daiquiri, West Indies Style .85	Old Fashioned (Scotch) .85	Jack Rose .70
Bacardi .70	Dubonnet .65	Old Fashioned (Bourbon) .75	Sazarac .90
Coffee .75	Rum .65	Perfect .65	Scarlet O'Hara .85
Orange Blossom .65	Moscow Mule .85	Pink Lady .75	Stinger or Side Car .90
Daiquiri or Martini .70		Grasshopper .90	Screw Driver .80

LONG DRINKS

Tom Collins a la Dempsey	.75
Rum Collins	.70
Cuba Libre	.75
Ward Eight	.85
Planter's Punch	.90
Singapore Sling	.90
Dempsey's Punch	.85
Rum Punch	.80
Sherry Cobbler	.65
Whiskey Egg Nogg	.90
Brandy Egg Nogg	1.10
Brandy Milk Punch	.95
Zombie	1.00
Jack Dempsey's Knockout	1.00
Bloody Mary	.95
Vodka Collins	.85
Sherry or Port Flip	.65

FIZZES and SOURS

Gin Buck	.65
Sloe Gin Fizz	.65
Rye, Gin or Rum Sour	.75
Scotch Sour	.85
Brandy Sour	.95
New Orleans Fizz	.95
Gin Rickey	.65
Golden Fizz	.85
Silver Fizz	.85
Gin and Tonic	.75

CORDIALS

	Pony	Drink
Domestic Cordials	.55	.75
Imported Cordials	.75	.95
Benedictine and Brandy	.75	.95

(FRAPPES 10c. EXTRA)

BRANDIES

	Pony	Drink
Remy Martin 3-Star	.75	1.00
Remy Martin V.S.O.P. (20 Years)	1.00	1.25
Courvoisier	.75	1.00
Metaxa	.75	1.00
Martell 3 Star	.75	1.00
Hennessy 3-Star	.75	1.00

RUM

Rum Carioca (Heavy)	.75
Rum Carioca	.65
Meyers Jamaica	.85
Bacardi White (Cuban)	.85
Bacardi Gold (Cuban)	.85
Bacardi (Puerto Rican)	.65
Ron Rico	.65

RYE and BOURBON BLENDED WHISKEY

Schenley Reserve	.65
Carstairs	.65
Calvert Reserve	.65
Three Feathers Reserve	.65
Imperial	.65
Hunter	.65
Seagram's "7" Crown	.70
Four Roses	.70
Park & Tilford Private Stock	.70
Schenley Canadian	.75
Seagram's V. O.	.75
Canadian Club	.75
MacNaughton's	.75
Early Times	.70
Jack Daniels	.90

BONDED WHISKEY

Old Overholt	.75
I. W. Harper	.75
Kentucky Tavern	.75
Old Forester	.75
Old Taylor	.75
Old Grand Dad	.75
Mount Vernon	.75
Old Fitzgerald	.75

SCOTCH WHISKEY

White Heather	.75
Black and White	.75
Vat "69"	.75
J. & B.	.75
Teacher's	.75
White Horse	.75
Dewar's White Label	.75
Haig & Haig 5 Star	.75
Cutty Sark	.75
Ballantine	.75
Martin's V. V. O.	.75
King's Ransom	.85
Johnnie Walker Black Label	.85
Haig & Haig Pinch Bottle	.85
Bell's (12 Years Old)	.95
Ambassador (12 Years Old)	.95
Chivas Regal (12 Years Old)	.95
Ambassador (25 Years Old)	1.25

MISCELLANEOUS

Southern Comfort	.85
Old Bushmills (9 Years Old)	.75
Laird's Applejack	.75
Rock and Rye	.65
Vodka	.75

SHERRY and PORT

Harvey's Bristol Cream	.85
Harvey's Bristol Milk	.75
Cockburn No. 25 Port	.55
Duff Gordon Sherry	.55
Imported Port	.55
Imported Sherry	.55
Domestic Sherry	.40
Domestic Port	.40
Burgundy or Sauterne	.50

DOMESTIC and IMPORTED CHAMPAGNE and SPARKLING BURGUNDY

Piper Heidsieck	12.00	6.50
Piper Heidsieck (nip)		2.50
Mumm's Cordon Rouge	12.00	6.50
Chauvenet Red Cap	9.50	4.75
Chateau Renault American Champagne	6.00	3.50
Chateau Renault	6.00	3.50
G. H. Mumm, Extra Dry	12.00	6.50
Moet and Chandon Dry	12.00	6.50
Veuve Clicquot Pansaroin Dry	12.00	6.50
B. & G. Macon, Royal Purple Burgundy	6.00	3.50

BOTTLED BEER and ALE

Budweiser	.50
Ballantine	.50
Dow Canadian Ale	.55
Heineken's Holland (Imported)	.65
Guinness Stout, Nip (Imp.)	.45
Bass Ale (Nip)	.40
Ballantine Ale	.50
Augustiner (Imported)	.70
Kalak Water	.35
Saratoga Water	.35
7-Up	.25
Coca Cola	.25

WINES

	Bottle	½ Bottle
Sauterne (Domestic)	2.50	1.35
Claret (Domestic)	2.50	1.35
Burgundy (Domestic)	2.50	1.35
Sauterne (Imported)	5.00	2.50
Claret (Imported)	5.00	2.50
Chianti Imported)	4.50	2.25
Burgundy (Imported)	5.00	2.50

Once when I was on one of my trips back from the ranch and going through Grand Central Station in New York, a lady I had befriended when she visited the ranch called and asked if I could arrange to spend a night or two in the City. She met me in downtown Manhattan and asked what I'd like to do. It was a Friday evening and at that time the Friday night fights in Madison Square Garden were a big deal. I told her that was where I want to start the evening. When we arrived we found the fights had been preempted by a Billy Graham Crusade. I said, "Well, I enjoy Billy Graham, so why don't we go." Later that night she said, "Kenneth, in case you accidentally meet my parents sometime, don't mention we went to see Billy Graham." When I asked why she said, "Because I'm Jewish, you idiot, and my mother is a typical Jewish mother." I wasn't sure what that meant, but assumed that mothers tend to be similar in all parts of the country.

My second year in the Catskills, we had a lot of accidents due to some unusual horses. I had picked out a good-looking small Palomino (about 10 hands). When it arrived on the truck the night before, I saddled it and rode it around the corral one turn. It reined great, had a smooth gait, plus it would be appealing to the dudes. The next morning when we were down at the corral looking at our new horses, I found out this Palomino would only turn one direction and had only one speed which was full speed ahead. Most of the time it would stand rigid and actually pose. I suspect it came out of a circus act someplace where its only duties were to run around the ring in one direction or pose for the audience. Regardless, I thought I could perhaps salvage something from my choice because he was so darn pretty. Jack Harris, another wrangler, was at the corral and he realized that a Paint horse he branded had a bad habit of rearing up about every thirty seconds. Outside the corral, we had a jeep trail going up the mountain that we would use as a runway for difficult horses. If you just let them go, by the time they reached the top they were a lot easier to handle. He and

I started up the mountain, and when we got almost to the top, his horse reared so he hit it between the ears with a large maple branch and killed it. Jack, who had a stiff leg from a gunshot wound, couldn't get out from under the horse. When it rolled over on him, it broke his leg in several places. I dismounted and cut his boot off and I remember pouring blood out from the bottom of the boot. It turned out his leg had been broken in sixteen places. I jumped on my trusty Palomino to head back for the ranch for help and it wouldn't move. I tied him to a tree, ran down and we got a jeep and came back for Jack. Another cowboy spotted my horse and saddle and decided was going to do me a favor and ride the Palomino back to the corral. It took off through the forest and knocked him off under a tree and put him in the hospital. My horse and saddle stayed lost for a week.

About a week later while breaking horses, one jumped the corral fence and rolled over on me breaking my collarbone. I remember getting up, almost passing out and sitting down on an anvil. Had I been about a foot closer to the anvil when the horse rolled over it might have been "all she wrote." About the time my shoulder began to heal, I broke it again in a rodeo. I had a cast on most of the summer. Jack and I roomed together since we were both nursing broken bones. We were well acquainted with the ranch nurse and she gave us a constant supply of morphine and Demerol plus a few other favors that helped us make it through the night. However, Jack's leg didn't heal properly and he went back to Alabama. I understand he had the leg amputated.

The good news part of this accident-prone story is that about six months after I'd returned from the Catskills and was back in school, I got a letter from the State of New York asking about my accident. It had not been reported to Workman's Compensation. I replied that I didn't know I was supposed to make a claim. I later found out the ranch had not reported some accidents because they were afraid of the effect it might have on their insurance. They said

I had to come back to New York for an examination and hearing which I claimed was impossible. They then supplied me with a doctor in Lincoln, Nebraska who, after an examination, assigned me to a physical therapist. He told me my shoulder was a mess, and besides the broken bones I had other damage to the joint. He did help me with several months of treatment. Eventually he finally told me that if I would do push-ups every day, my shoulder would probably be okay. I asked him if 100 a day would be adequate and he said "yes." Since that time I've done 100 pushups almost every morning. Approximately one year later I got a letter from New York's Workman's Comp telling me they determined I had 25% permanent disability to my left shoulder and they sent me a check for $1,543. At that time, I was on a $200 per month fellowship and Pat, who I had married three years earlier, was making $350 a month as a lab technician. I remember we looked at the check and said this is "manna from heaven." It sure helped make graduate school easier. In the end, I was classified as 4F, and even though I can't throw a football or baseball or lariat, I never could anyway so I guess you'd have to call it a "lucky break."

–K–

Two Girls and a Mare

K. S. Eng

I wrangled dudes in the summer
I didn't have a worry or care
My favorite girls were Piney and Dixie
And Tilley was my favorite mare

I spent the daytime with Tilley
Wrangling children and the guests
Nighttime was spent with the ladies
Piney and Dixie topped my list

I had a bad habit of mixing their names
Both the girls and my filly
Confusing Dixie and Piney
meant trouble
And neither liked to be called Tilley

Dixie and Piney left the same week
Said I wasn't very stable or smart
I chalked it up to experience
I was sad, but no broken heart

The next day I sat slumped
in the saddle
As Tilley drank from the water trough
First I called her Piney then Dixie
That's when Tilley bucked me off

I learned a lot that summer
Dixie and Piney were a great pair
But I switched to girls with
simple names
And rode a gelding instead of a mare

-CHAPTER 4-

Graduate School

After finishing my Bachelor's degree, I decided I definitely liked animals better than plants and I needed to spend more time in the Animal Nutrition field. I had several university teaching and research fellowship offers but I chose to stay at the University of Nebraska for my Master's degree and then went to Oklahoma State University. Working on my Master's I was fortunate to have Dr. John **Matsushima** as my major advisor and he was a great mentor. He and his wife, Dorothy, were also good friends—and graduate students never have too many friends. I took several nutrition classes and more science classes, and I managed to complete my degree and research in three semesters. Some of the professors said I seemed to be in a big hurry and I told them, "You got that right."

On weekends several of us would compete in small town rodeos. Also our college rodeo team finally got off the University "black list," and we scheduled a spring rodeo at the Coliseum. In conjunction with this we decided to have a buffalo barbeque—and that's where it got interesting. I don't recall where we got the buffalo, but they let us take it to the meats lab to be slaughtered and processed. We ran it into the kill chute and hit it a couple of times with the stun gun which did no good whatsoever. One of the team members said, "I've got an old Winchester in the pickup." If I recall correctly, it was a .38-55. He shot the buffalo and put him down, but when we opened the chute gate, and I tried to put the leg shackles on him, he got up, charged around, and tore up everything in sight. A meats lab employee who was a big burly fellow kept yelling at us, "Don't shoot; don't shoot," as he chased the buffalo with a sledge hammer. Finally, everybody including the buffalo collapsed in a heap, and we got the job done but the end result was the rodeo team was once again on that "S" list.

By the time I reached Oklahoma State University, I was pretty burned out on college. but I thought I had too much invested to quit without a Ph.D. I ended up with a major in Ruminant Nutrition and minors in Physiology and Biochemistry. Quite a departure for a fellow that five or six years previously had not even entertained the idea of going to college. Nonetheless, I resolved to work hard and get out as soon as possible, not only because of my boredom but also because Pat and I had one little girl, Laura, and another one on the way, Julie.

Regarding my studies, the first thing I decided to do was to get my language requirement out of the way. I had a choice of either German or French. Since I knew a little scientific Greek and Latin, I chose French and proceeded to audit a French class and study like hell. If you had an approved textbook to translate you could "test out" at any time. A month later, I took my French exam, got lucky and passed it and then started to work on my other classes along with my research.

We had a lot of good graduate students at Oklahoma State who later did well in their careers. Even so, some of the professors were not easy to get along with. I was also somewhat unhappy because of my lack of control over my own destiny. There were several professors who really didn't like each other, and if you were unfortunate and got a couple of them on your Graduate Committee. it spelled trouble. My research went fairly well, and writing had always been easy for me, so I submitted my thesis in record time. That's when my troubles began. I believe there were five people on my committee, and they would pass my thesis back and forth—each making corrections on corrections on top of corrections. When I would subsequently re-edit it, one of the other committee members would make additional corrections or want it changed back to the original. Fortunately, one of my committee members was Dr. W.D. Gallup who was a former editor of the *Journal of Animal Science* and renowned for his work as a biochemist as well as his writing ability. I had an office next to him in the biochemistry lab. At this time he had been ill for over a year. I finally caught him in his office one day and asked if he would have time to read my thesis. He did, and he really "tore it up" but in the end it read much better, and after I made the corrections he suggested he signed it on the spot. After he signed it, the rest of the committee fell in step. They said the same thing at Oklahoma State as the University of Nebraska—"You sure are in a hurry." And I again said, "You don't know how right you are."

In fairness, I'm sure I was a "pain in the butt" and didn't fit the accepted mode of a Ph.D. graduate student. I've always been an independent thinker, and while I can accept authority, that doesn't prevent me from questioning it. Also, being in a hurry doesn't necessarily fit well with some college professors. Finally, most of the professors where veterans of World War II (many distinguished) and they didn't have a lot of tolerance for a wet-nosed kid who was already accustomed to making his own decisions. Looking back, most of my professors were pretty good people (especially my major advisor, Dr. Nelson), but the fact that many of them didn't get along well made it difficult for grad students.

I had my class work done, thesis approved and orals passed prior to my twenty-fifth birthday. There was only one problem—no jobs were available. Most of the Ph.D. graduates were accepting "post-doctorate positions," but I swore I would never do that. I even considered going back to horseshoeing. However, at the last minute I got a job offer at the Texas A&M Gulf Coast Experimental Station and accepted it before they could change their mind. It turned out to be a great job, it paid a whopping $7,200 per year, and we were on our way to Aggie land and the Gulf Coast with a new Ph.D. and a real job.

-CHAPTER 5-

Texas A&M—Aggie Days

The Texas A&M Gulf Coast Experiment Station was located near Angleton, Texas approximately forty miles south of Houston and fifteen miles from the Gulf. The Station Superintendent was J. C. Smith and the pasture research man was Marvin Riewe, both of whom had an agronomy background. I was to work on nutrition research with the cow herds which consisted of a pure bred Hereford herd, a pure bred Brahma herd, and we were developing a F1 Braford herd. Texas A&M was at the forefront of much of the F1 research that was interesting. Riewe and I were involved in a joint Southeast research project involving almost all the USDA and southern universities, and we were doing research on coastal Bermuda which was a new hybrid Bermuda. This was advanced and intriguing research, but I also had a lot of other ideas Luckily, I was given completely free rein as long as I could find the additional money needed. We ended up with a pretty nice feeding facility, metabolism facility and a reasonably good nutrition lab.

I also built some climate control feeding pens for lambs with the goal of studying the interaction between temperature and diet fiber level on intake and performance. To study the difference in diet density and fiber digestion and the effect of heat increment on intake at high temperatures, I procured a granulated polyethylene material from a nearby Dow Chemical plant. I compared diets with approximately 30% granulated polyethylene to diets with a high roughage or crude fiber level. The results were thought-provoking, and I consequently published a couple of papers receiving much curiosity from scientists in various parts of the county. One was a nutritionist named Dr. White from an Louisiana State University experiment station. He wanted to know where to get polyethylene. I knew there were several chemical plants producing polyethylene around the New Orleans area, and suggested he check with them. He started a feeding study using steers. About a month later he called saying, "This isn't working." The polyethylene cattle were getting a bigger gut, lower intake and thinner as the days went on. I never encountered that problem but was interested in his observation. I drove over to see Dr. White and found that his polyethylene source was not granulated but cubes of about ½ to 1 inch in diameter. I told him, "I don't think these cubes would pass through the rumen, and if you check the rumens, I bet you they're full of polyethylene." Sure enough that was what had happened. We all had a good laugh and I would have forgotten about it, but nutritionists from Farmland Industries and Dow Chemical began calling me asking for more information. Based on these observations and data and a little of their own research they obtained a patent on a product called "Roughtabs" which involved the feeding of polyethylene cubes for the first few weeks of the feeding period. You then withdrew them plus any other roughage with the assumption these cubes would recycle in the rumen and provide the necessary roughage stimulation in the rumen. It was a great concept and they did get patents and feeding approval; however, when they took the program to commercial feeding operations, it never worked as planned.

I worked on another piece of original research that involved ammoniated rice hulls. The product was called Deltrix-10 because the protein equivalent was approximately 10-11%. We conducted several trials with it and found it was a reasonably decent low-quality roughage substitute in high concentrate diets. The parent company had a fair amount of success marketing it into the feed industry for several years. More interesting than the product itself were the individuals involved who were venture capital types of people and ranch and oil people from the Houston and South Texas area. The man who developed the patent was the prototype of a mad German scientist. To call him eccentric was a gross understatement but he was dynamic and weird enough to attract a lot of investors. Rice is, of course, a big crop in the Gulf Coast area, and rice hulls are as close to totally indigestible as any natural occurring feedstuffs. Thus there had been a big disposal problem with them. Rice hulls are composed of approximately 40% crude fiber, 20% lignin, plus 20% silica and in their native state are truly indigestible. The process involved ammoniating the rice hulls under high pressure in a large reactor. The venture capital company also worked at ammoniating a lot of other products and looked at additional treatments of fiber with calcium hydroxide, sodium hydroxide, and other materials.

There were several impressive people involved either from a technical standpoint, or as investors or promoters. These included Harry Konen from Houston, who was a poultry nutritionist, Forrest Warren, who was an independent Houston oil man and rancher, some personnel from King Ranch, and several people involved in the Weingarten food stores. Konen was a great guy, a chemist by trade and a very successful, independent poultry nutritionist at that time. He was my first exposure to independent consultants, and when I did a little part-time work for him I "caught the bug." Warren, who I will discuss more later, became a great friend. He died early, but he was involved in several feedlots in the Texas Panhandle area, pasture cattle throughout Texas and a ranch near the edge of the Crockett National Forest. I helped him design a feeding research station at his East Texas ranch, and we conducted several feeding studies there. My first consultant fee was $5 per hour plus expenses and I was worried I might be overcharging him. One little-known fascinating aspect of ammoniating products, especially those high in soluble carbohydrates, is that there is occasionally a toxin formed that affects an animal's nervous system. In the mid-1950s, several molasses companies worked with an ammoniated molasses product.

The reason they discontinued the practice was the occurrence of what was referred to as the "Crazy Cow" or "Mad Cow" syndrome. Some cattle would go berserk and charge into fences, cowboys, and so on, but they would usually recover. I observed this occasionally in our research facility. You would also sometimes see it in calves that were nursing cows who were fed ammoniated products. This correlation was later confirmed by Wisconsin researchers using sheep as experimental animals. They attributed it to an alkaloid formed in the ammonization process. One weekend an animal in one of the experimental pens I was walking through began to shake and get that wild-eyed look. It charged and ran over me then hit the fence and collapsed. Five minutes later it was back to normal but it was an exciting and painful observation.

The eccentric scientist convinced the investors they should put a plant in the rice-growing areas in Mexico. He was sure that with the proper equipment he could produce a 20% protein product. The Weingarten organization backed this project along with some Mexicans in the entertainment business who knew nothing about cattle but did know about promotion. However, the product they began producing not far from Veracruz would never pass lab tests conducted by the Rockefeller Foundation. I knew Dr. Ned Raun who was head of the Rockefeller Foundation in Mexico City, and they asked me to help convince him to register the product and also to visit the facility near Veracruz. First I went to the plant which was near a small town called Cordova, Mexico on the edge of the jungle. It was a beautiful area but a strange place for the plant. They made a "run" of their ammoniated material for me to examine. I could tell by looking at it that it was not going to be that high in protein but, when they ran it through the lab he had designed, it checked out at exactly 20% protein equivalent using the Kjeldahl method. I asked them if they ever did a water blank to check their reagents and they said, "What's that?" I ran a water sample through the Kjeldahl procedure and guess what: it tested 10% protein.

Obviously their reagents had been "doctored." This solved the mystery of why the product wasn't passing specifications with the Mexican authorities, but they still wanted me to meet with Dr. Ned Raun and the Rockefeller Foundation. They were convinced that maybe with some *mordida* (bribe), this could be resolved. They arranged an elaborate dinner with Raun and I and their investors. I finally convinced them that their product was actually 10% protein rather than 20% protein and money would not change this.

I returned one more time to reconfirm these observations and have another round of parties. After it was all over, one funny comment the elder Mr. Weingarten made was that he should have known better than to invest in anything that "needed feeding or ate at night." I also got to be a good friend with other Weingarten people. One of them was an accountant from the East who had previously been involved in the entertainment business with Mercury records, Playboy, and the like. He had married a blond Texas girl quite a bit younger who was in some ways a stereotype blond. Once we were with him and ran into Joe Louis at the Dallas airport. The Weingarten man who had known Joe from his entertainment business days introduced Joe Louis to his wife who promptly asked him, "And what do you do, Mr. Louis?"

While at the research station I would go to the main campus once a month and became well acquainted with Dr. Butler, Department Head of Animal Science, and Professor Riggs who are both great people. O. D. Butler was a great department head and Riggs, while not flashy, was a classy fellow. Butler was a World War II veteran and an officer in the Rangers. He was fair but could be abrupt. At one of the annual beef conferences, Bill Ellis, who was a new staff member, said he had a friend from Rowett Institute where he done post doctoral work who he could secure as the main speaker for the evening Beef Conference dinner. This gentleman was Dr. Reg Preston "not Rod Preston," and he came dressed in slippers and blue jeans. The next day he flew from Houston to Cuba to head up

Castro's Agricultural Research. This really embarrassed Butler and I remember him telling Ellis that the next time he recommended a speaker, he damn well better check his politics. Another amusing Butler incident was a conversation he had with Bill Pope who was then Associate Ag Director at A&M. They were just completing a trial at the Beeville Experiment Station comparing steers versus bulls started as calf feds. As expected the bull calves outperformed the steers but Pope told Butler that he sure wouldn't want to eat bull meat. Since some of the carcasses would be processed at the A&M Meat Research Lab, Butler suggested he buy a bull and put it in his locker, and Pope could buy a steer and they would compare them. Several months later Bill commented to Butler that his steer was really good eating and asked how was the bull? Butler who was in charge of the carcasses replied to Pope, "How do you know you got the steer?"

After three years, I decided I wanted to go into the commercial industry even though Texas A&M asked if I would transfer to the main campus. I experienced some remorse and left a lot of good friends with whom I would associate again around 1970.

A couple of footnotes to the ammoniated rice hulls story: an Ag writer from the *Houston Post* interviewed me about our research, and wrote an article including the amount of rice hulls to be potentially processed. It amounted to more than twice the rice hulls produced in the entire world. When I called him and pointed out the mistake, he said, "Son, don't worry about it, you got your name in the paper and it's just a bunch of zeroes anyway." Also, about ten years later the German scientist and some investors in the Philippines called me to do some consulting work on a new project designed to grow beef from tissue cultures to feed all the Philippines Islands. I said I was too busy. but wished them luck.

-CHAPTER 6-

Leaving A&M for the Corporate World

Though I loved the Texas A&M University people and my job, my association with the business community in the Houston area convinced me I wanted to go into industry. In 1965 it became obvious a small group of independent nutrition consultants, mainly on the West Coast, were having a huge impact on the emerging commercial feedlot business. Ralston-Purina offered me a job as feedlot consultant for the western part of the United States. At that time, they had only one other company consultant and he was in poultry. I didn't want to go to St. Louis, but they told me I could relocate to either their regional office in Denver or Sacramento. So I went to Denver first, then moved on to Sacramento. I stayed with Purina three years and learned a lot about the business aspects of a large corporation. At that time, Purina was the dominant feed company in the entire world, and they were anxious to "tap" the consulting business. However, I found that representing a feedlot client *and* the Purina Corporation simultaneously and fairly was a difficult challenge. It was also obvious when I went back to St. Louis once a month that there was a lot of internal corporate competition for a piece of the consulting action. Nonetheless, it was a company full of smart, dedicated people, and a young person such as myself couldn't help but learn a lot.

Dr. Dean Hodge left the Kern County Cattle Company and took a job as head of beef research in St. Louis about the same time as I came onboard at Purina. About two years later, they added several other consultants whom they based out of St. Louis—and they were a pretty illustrious group. Most of them later moved into the independent consulting business and included people like Jack Martin, Mel Karr, Les Kuhlman, and others.

In 1968 the company asked me to relocate to St. Louis and to take on a job as liaison between the sales and marketing group, the computer group, the research and consulting group. I thought that would an impossible task and besides, I had vowed never to live in cold country again so St. Louis was off limits to me.

As I was leaving, Bill Lane who was Director of Purina's sales group said he could see I chose to become a big fish in a little pond rather than a big fish in a big pond like I could have been if I stayed with them. He was smart, super-aggressive and he may have been right, but I doubt it.

-CHAPTER 7-

Early Independent Consulting and Consultants

In the summer of 1968, I decided to hang out my "Independent Consultant" shingle when Dr. Bart Cardon of Tucson, Arizona contacted me. He had small commercial feed company and research and consulting group in Tucson called the Early Fat Feed Company. Bart was an incredibly impressive individual—large with a great voice and he could give a hell of a speech. He was in the process of merging or buying a larger feed company in Arizona called Arizona Milling, but he wanted to stay in consulting. He suggested we pool our talents. That sounded okay with me and I moved to Casa Grande, Arizona where I stayed off and on for the next fifteen years. My consulting business grew rapidly, and at the end of the first year, it was both busy and profitable.

However, Bart and I had different visions of the consulting group. I wanted the consulting to be separate from his company, but Arizona Feeds could own a portion of it. Bart wanted the consulting business to be part of Arizona Feeds, and in turn I'd get partial ownership in that company. Early on I suspected that Arizona Feeds was going to encounter problems—which it did. I told him, "Bart, why would I want part of your company when it's unprofitable and my consulting business is profitable." He said, "I guess you wouldn't," and we had an amicable parting. I promised I wouldn't pursue Arizona accounts but the rest of the consulting world was mine if I could handle it.

Again, I learned some new things about business when I was associated with Bart. One was don't fall in love with your own voice and ego if you want to develop a successful and profitable company. As I mentioned previously, Bart had a great speaking voice and presence and was perhaps the best speaker I have ever met. However, he would travel all over the world and sometimes not even charge expenses just to be on a program. I learned early on that if you think you are a good speaker and have something worth saying, charge for the effort. If they are unwilling to pay, your information must not be very valuable. Arizona Feeds eventually went broke, and I believe Cargill owns it. Bart later became Dean of Agriculture for the University of Arizona. It was a job he was well suited for and where he ended his career on a high note.

Bart and I remained friends over the years. As smart and impressive as he was, it was obvious that managing a successful corporation based on a personality cult was difficult. That's a good lesson for all of us, but as Mac Davis says, "Lord, it's hard to be humble."

Beginning in 1969, I was totally on my own and without partners. Over the years I've had several individuals ask if I would consider taking on a partner and lots of them were very talented such as Dale Furr. However, I didn't think it would be fair to them or me. If we had a client or a project that was fascinating enough to take on I'd probably want to meddle in it. Over the years, I worked closely with several consultants but it's been on a pay as you go, and not a partnership basis. I suspect this limited my growth, but I never thought bigger was better. One way or the other, I had more work than a rational person could handle and I suspect that at one time during the 1970s I represented more clients with more cattle than any individual in the country. There was lots of work available, not much competition and I was still young, ambitious and had the curse of a curious mind. I'm not necessarily bragging about this because there was a personal cost involved. During that time, I didn't mean to be unkind, but I occasionally neglected family and friends and others who were important in my personal life. Some might say that was ego, but it really wasn't. I just wanted to see what was around the next corner or over the next mountain. Again, the curse of a curious mind.

The independent feedlot consulting business started on the West Coast around 1960 and it must have been a good business model because it survived and thrived for fifty years. When I joined the ranks in 1968 there were approximately seven active and successful independent consultants. They included Jack Algeo, Gene Erwin, Jim Nofsiger, J. D. Autry, Jim Elam and Ross Cooley. Cooley lived in Amarillo, Erwin in Arizona and the rest were Californians. Bill Roberts (hired by Erwin) and I were the last to join the original group. Each of us were different, although we also had many similarities. Many had airplanes and research facilities, and because there was abundant business available, we stayed out of each other's hair. Jim Elam and I are the only ones of the original group still alive.

Many practices now taken for granted were pioneered by this small group and our clients. This would include high concentrate rations, the addition of feed fat, higher levels of total ration protein, NPN, etc. We also pioneered many byproducts such as citrus pulp, almond hulls, potato waste, etc. You'd be surprised at the number of "experts" at that time including several University people who believed steam flaking did not pay, 10% protein was the max required in a finish diet, feed fat was not a suitable ingredient and you needed at least a 12-15% minimum finish ration roughage level. In the early '70s, a marketing group who had done a survey told me that approximately two-thirds of the cattle fed in the United States used programs formulated by this small group. Today I suspect only 30-40 percent of the feedlot cattle are served by truly independent consultants.

In the 1970s a number of new consultants joined the industry including Wally Koers, Jack Martin, Mel Karr, Dale Furr, Steve Armbruster. Their entry was followed by several individuals who are still active including Abe Turgeon, Bill Dicke and his group, Dave McClellan, Jim Simpson, and Mark Miller. Hollis Klett took over Gordon Reiner's Extra Factors business and has since added nutritionists including Keith Hansen, Nathan Elam, Bob Beville, Gary Holcomb, Scott Schake, etc. Their business model is a little different but it must work. In addition, most commercial companies now have company consultants who work with their customers rather than salesmen. This was a business model pioneered by Ralston Purina in the mid '60s. Currently, Midwest PMS is a successful feed company with a number of consultants.

Graduate student groups often ask me what does it take and how can I get started in the consulting business. It's a complicated question but among the requirements is a curious mind, a flexible mind, a high-energy level and the ability to make quick decisions based on what you believe is the best for your clients.

I'm often asked to compare today's consultants with the original group. This is also difficult because there have been so many changes in the industry and society in general. When I started there was less competition between consultants and more friendship and camaraderie than you find today. Also, early consultants were more likely to think outside the box, making quick changes if they were in the best interest of their clients even if they weren't particularly popular. I now hear more conversations and concerns about such things as losing clients, legal liabilities, and reluctance of clients to make changes than I heard in the past. Some of this may be due to the changes in the makeup of clients and in the industry in general. One of the great advantages I had is that I never worked for someone I didn't like. I tell my small group of employees that their biggest fringe benefit is we only do business with good people that we truly enjoy. Life is too short to drink cheap wine or associate with negative people.

Dale Furr, a successful consultant from Hereford, Texas, was a very close personal and business friend for almost fifty years. We met in 1960 at Oklahoma State University where we were both working on Ph.D.'s. Dale retired from consulting approximately ten years ago and died three years ago. In between, we worked and played together, and perhaps more important, had numerous twelve- to twenty-four hour gin and domino games. We also did a lot of fishing. Dale was extremely successful consulting mainly in the Texas Panhandle area. He had a number of great clients. Most people thought of him as a very conservative quiet person, but there was definitely another side. He was an intense gambler. Our gin and domino games resulted in approximately a half million dollars of wins and losses between the two of us over forty years. In the end, we came out about equal, but he was an intense gambler whether it was between the two of us or he was in Las Vegas. When he got out of town he also had a wild side. Bob Beville, who worked with him, and I always said that when Dale uttered the phrase "look out south Dallas" you knew it was time to run for cover.

Although we were close friends, we were also very different. Caroline called us the odd couple. I was Oscar and Dale was Felix. Dale was a complex and fastidious person. Many would be surprised to learn he was also stubborn and filled with self-doubt.

He always worried about what other people thought even though he was very successful. He would often tell me he wished he had my attitude and was as successful. I would reply, "Dale, you're really successful and quit sweating the small stuff." In spite of our differences, we had great times together including fishing and gambling. On one of our nutrition trips to Bodega Bay north of San Francisco, there was a large group including Jim Mickus from Cargill who loved to play gin even though he wasn't very good at it. Dale and I would argue over who got to play him, but our plans ended badly because the ocean was really rough. Almost everyone, and especially Jim Mickus, got really sick. The morning we were to start fishing it was rough so everyone went to the restaurant for a big breakfast which was a mistake. On the way back to the boat a seagull flew over and crapped on Mickus's head. My friend, Billie, was with me and in her broken English said, "It is very bad luck when a bird shits your head." She went back to the cabin rather than go fishing. To make matters worse, Keith Hanson caught the only fish, and I believe he got sick immediately thereafter. It could have been worse because the following week the same boat capsized and one person drowned.

Billie and Dale were good friends and she always worried about Dale because of his diet and also because he didn't have windows in his office. After Dale had a couple of bypass operations he became fanatical about his cholesterol levels and got them extremely low. I tried to convince him this wasn't good for him mentally, and Billie always said if he didn't get more sunlight, he would lose his mind. I don't know if it's related, but unfortunately Dale did extend his life after his heart attacks only to suffer from Alzheimer's the last four or five years of his life. One good thing he did following his bypass operation was to quit smoking after being a long-time chain smoker. I told him that his heart attacks and swearing off smoking added at least twenty years to my life because of the secondhand smoke I experienced during our gambling sessions. Bottom line, I miss him and regret that his final years ended as they did, even though his wife Jan took good care of him.

-CHAPTER 8-

Return to A&M—Part-Time Teacher

In 1969 Dr. O. D. Butler, A&M Department Head of Animal Science contacted me and asked if I would design and lecture the first class they planned to offer in the Feedlot Management Master's Degree Program. It was to be a three-hour credit graduate level course, and he said I could schedule the lecture times so they wouldn't interfere with my consulting. We agreed on my fee and I scheduled it for 7:30 AM to 9:00 AM on Monday, Tuesday and Wednesday mornings. The problem of travel is never simple to and from College Station, Texas. I finally settled on a routine of flying in and out of Austin, Texas and driving a car from Austin to College Station. Because of the travel, it turned out to be more work than I anticipated, but it was enjoyable and fulfilling.

Dr. Lowell Schake was on the A&M Staff had done a lot of the ground work on the overall Master's Degree program so I was spared most of the paperwork and bureaucracy. Lowell later became Department Head of Animal Science at Texas Tech and I believe he is now retired in Corpus Christi, Texas and writing books.

I was blessed with a lot of good graduate students in these classes. At the time, A&M had the student's rate classes for quality and content, and my class was fortunate to get top ratings both semesters I lectured. In later years, Dr. Bill Messe taught the class, and I understand that due to his efforts, it continued to be very popular.

The second semester I taught at A&M, I kept a room at the Ramada Inn on the corner of the campus and Highway 6. At that time the Aggies didn't have a good football team, and the joke was the sign out front that said Highway 6 – Texas A&M 0. Times have definitely changed. I liked the Ramada Inn location because it was close to campus, had one of the few private clubs in town, the Camelot Club, and the Bank of A&M was across the street. Caroline was working at the Bank of A&M and I first met her one day when cashing a check. The airline had lost my luggage, and I was to give a talk later that night. I asked her where I might buy some clothes. She said, "It's almost bank closing time, and if you'd like. I'll take you to a clothing store." It was an invitation I couldn't refuse.

We became good friends, and when my teaching appointment was over she said, "Kenneth, someday I am going to marry you." Like John Wayne I said, "That'll be the day." I didn't see her for almost two decades. Eighteen years later when I was single, she called from Houston, and after a brief conversation asked what I was doing. I said, "Not much. There's a late flight to Houston Hobby, and I'll be on it." She met me at the gate and it was a storybook beginning to a twenty-two-year ride of a lifetime.

-CHAPTER 9-

'70s Consulting—Texas Style

The cattle feeding industry in Texas, especially the Panhandle, began growing in the early 1960s, then experienced spectacular growth in the late '60s and early '70s. Much of this was fueled by an **entrepreneuria**l spirit and "can do" attitude. Perhaps the biggest driver was the advent of hybrid grain sorghum in the late '50s. This resulted in surplus grain sorghum or milo in many parts of Texas. Before this time, milo was often shipped to the Panhandle with the price determined by the Panhandle price less freight from wherever it was produced. Other factors included a strong Texas Cattle Feeders Association located in Amarillo, and the advent of tax and investor feeding. Also there were lots of Texas and Southeastern calves and wheat pasture yearlings. Keep in mind that distance hauling grain or calves was not a big item at that time due to low fuel prices. It was easy to put wheels under your commodity and ship it long distances.

In many ways, the hey-day of the Texas and Southern Plains cattle feeding industry was a boom time rivaling that of the gold rush in California and Alaska and the early oil industry in Texas and Oklahoma. One of my favorite George Strait songs is "Amarillo by Morning," but "Amarillo by Evening" during that time was more exciting. Every commercial plane to Amarillo was full of cattle people or those in allied industries. The motels, restaurants and clubs were full. It was not unusual for large business transactions to be consummated on a restaurant or club napkin. Many of those deals hung together better than some of the later sophisticated ones. It has been my experience that if you use good judgment, a cattleman's word is his bond, but beware of people who require complicated contracts and everything signed in triplicate.

Time, space and poor memory don't allow me to cover all of my clients in this book. Many of the events I discussed are the more exciting and humorous ones. This doesn't mean the business was all excitement and humor, but I'll spare you from most of the negative or boring stories.

One of my first clients was Producers Grain Corporation (PGC) of Amarillo, Texas. They were the world's largest buyer and handler of milo, and they were interested in feedlots not only as investment but also as a way to expand milo usage. They first built a 25,000- to 30,000-head feed yard west of Plainview, Texas in '69 or '70, and later another one approximately thirty miles west of Amarillo. Moody Taylor was their manager. At one time he had managed both south and west Texas feed yards and also traded a lot of cattle. Moody was a clever and sometimes mysterious fellow. Regardless, he was the kind of person you would want on your side if you got in real trouble. PGC also had a feed company, and they wanted me to hire a nutritionist for their feed division that could eventually work into the feedyard nutrition business. I suggested Dr. Ray Hinders and they subsequently had a long association until Ray decided to move to California and take up independent dairy consulting. PGC also planned a partnership with the Shah of Iran on an Iranian feed yard, but the project never got off the ground. Just before it was planned to begin, the Shah was thrown out and the Ayatollah had other interests. No one wanted to do business with him anyway. Originally, they wanted me to go to Iran on the preliminary project. I refused because it was a one-month trip and I didn't have the time to spare. I told the PGC man in charge that I wouldn't be going, and he said, "You have to go because we've already put your name in, and they've accepted it." I replied, "Why don't you send Ray Hinders because he's your employee." We finally agreed that I would fill out the application required, then when it was time to go, I would be ill and they would send Ray in my place. I filled out a complicated application and then signed another form and asked their Amarillo secretary to type and submit it. When I came back to PGC a month later the Iranian project leader met me at the door and saying, "Who the hell do you think you are, nobody gets that kind of consulting fee." I had asked for $500 per day plus expenses and the secretary mistakenly typed in $5000 per day. In the end it didn't make any difference because the project crashed when the Shah was overthrown.

Another early client who built a feedyard about forty miles west of Lubbock, Texas was Wayne Owen. The feedyard was named Owen Brothers. Owen Brothers had originally been horse and mule traders in the Arkansas and Missouri area. They were said to be the biggest supplier of horses and mules to the Army during WWII and thereafter, they were used when mules were needed in countries such as Greece who were fighting the communists. An acquaintance who accompanied some of the mules from the U.S. to Greece said the biggest problem with the program was that many were eaten by the starving Greeks shortly after they got off the ship.

Wayne had a big heart, but he was also extremely independent along with a volatile temper and colorful language. Furthermore, he had the soul of a trader, and whether it was the futures, Las Vegas, or trading cattle, he was "all in." Nonetheless, he ran a successful feedyard for several years until bad markets and personal health problems won out.

Wayne developed an extreme dislike for most packers and packer buyers. His manager, Ed Bricker, tells a funny story about how he, Wayne and an Excel buyer were driving around the feedlot in a snowstorm looking at fat cattle. Wayne always wanted to drive his Lincoln, and when they got to the far side of the feedyard, about a mile from the office, Wayne told the buyer to look at a pen of cattle. The buyer said they didn't look fat, but Wayne insisted he get out and walk the pen. When he did, Wayne drove back to the office and left him to walk back in the snowstorm. Ed told Wayne he thought he should go pick him up, but Wayne said if he did, he was fired so the buyer walked.

Another illustration of Wayne's volatile personality is a time when he called and insisted I meet him at his office which was on the top floor of the bank building in downtown Lubbock. He was with two feeding customers who were associated with a packing plant back east. They were unhappy with the death loss in cattle they were feeding with Wayne.

I said, "Wayne, why do you want me when it's a death loss problem? Get your veterinarian."

Wayne explained, "You're tougher than the vet, and besides, the vet can kill more cattle than IBP."

We met in his top floor office, and the discussion got hostile in a hurry. Finally, Wayne motioned to me and said, "Let's you and I talk in the outer office." He instructed me, "Eng, go back in and settle with those guys if they offer something that is remotely fair. But if they don't, tell them when I come back and find they're still in the office I'll throw one or both of them out the window." I went back in and told them the situation and suggested that we settle and they leave pronto. They did.

I believe my first client in Texas was Gene Newman who owned Veribest Cattle Feeders east of San Angelo. He and Johnny Bonner, who was in the cattle and feed business in San Angelo, were best friends. Gene later hired Johnny's Aggie veterinarian son, Wes, as manager. Gene was a good cattle and horse man who readily chased new ideas and new horizons. He was on the nervous side, and his friends said he'd wear his Levi's out from the inside.

Gene was one of the first people to bring light calves in from the Carolina's. This meant that Wes stayed very busy. Gene had also read about the short scrotum method of castration which is really not castration but pushing the testicles up into abdominal cavity and banding the scrotum. The end result is that they look like they are castrated but, there was still a lot of testosterone activity. Gene tried it on one set of calves in 1973, and they fed well and sold with no discount. Gene had just bought a large ranch in Montana, and he told me that he wanted to short scrotum all of his bull calves. Wes and I tried to talk him out of it. but when Gene set his mind to something he was hard to change. He shipped 2,500 Montana short scrotum bull calves to the Texas Panhandle wheat pasture. The first thing that happened is the short scrotum bulls got real active on wheat pasture and there was so much bulling and riding that he shipped them to the feedyard. At the feedyard, the fat cattle buyers informed him that they would definitely not be in the market for those cattle because they were bulls in disguise. Gene finally gave up on the project and Wes had to remove the testicles from 2500 short scrotum bulls. Wes said that was the hardest he ever worked.

On the subject of Wes and castration, Gene sent a Simmental yearling bull on a load of cattle to the feedlot. He had left a message for Wes not to cut the bull because he wanted him for a herd sire. Wes didn't get the message and when Gene arrived and asked Wes if he had noticed the Simmental bull on the load. Wes said, "Yes, and I castrated him off the truck."

Wes Bonner stories are abundant and legendary. My favorite personal Bonner story is when Toad Tucker came by the feedlot one afternoon when I was there. Toad asked Wes if he could go deer hunting on Wes's ranch. Wes said that would be fine, and Toad said, "Eng, you just as well come with me, and we will meet Wes at his ranch house and celebrate our deer kill later that evening." The hunting wasn't particularly good, but I shot a small buck about twilight, and Toad and I headed back to the ranch house. To say the house was primitive would be complimentary. Among other things, there was a den of rattlesnakes and skunks under the front porch, and when you walked up the porch they let you know you were disturbing them. We skinned the deer, poured some whiskey and were sitting at the kitchen table discussing the big hunt. Wes's wife-to-be wasn't home, but her African Grey parrot was there in a cage. Every time Wes would walk by, the parrot would go crazy and try to attack Wes. Wes said the bird acted up because he had removed its infected claw, and the parrot subsequently never liked him. Toad and I thought the bird, whose name was "Bird," needed an attitude adjustment so we offered it whiskey. Believe it or not, that parrot loved whiskey and soon it became mellow, hanging upside down from its perch and talking up a storm. About that time, Wes's lady friend came in, said hello and then walked up to the cage and said, "Hi, Bird." Her first clue that something was amiss was that the bird was talking ninety miles a minute and hanging upside down from its perch.

She started yelling, "You've given my bird whiskey. This might kill him, and African Greys are very rare and expensive. You're just stupid SOB's."

Toad and I exited quickly leaving Wes to sort things out. You wouldn't have thought things could have gotten worse but I believe later that weekend the Texas Cattle Feeders Association was having their annual meetings, and Wes may have been President. The night before they were to leave for the meetings, skunks fell through the closet ceiling and sprayed all their clothes. This at least got the lady's mind off her parrot.

Gene later sold the feedlot to Wes and some neighbors, and Gene concentrated on Montana ranches and cutting horses. Gene was a good horseman, and really got into cutting horses. They said his main problem was he tried to buy cutting horses like calves, a truckload at a time. Gene died unexpectedly of a heart attack in his early fifties. He was staying at a hotel near a cutting horse contest in Austin, Texas. I was at a conference in Alberta, Canada when his wife called me at 5:00 AM that morning breaking the news to me. I don't suppose there is a perfect time to die, but this surely wasn't it. The cattle, cutting horses, land business and most other things in agriculture were a wreck. Gene had an insurance policy for Johnny Bonner in compensation for him to be executor of his will. To illustrate what a good friend of Gene's and the family Johnny was, he signed the policy over to the family and didn't take a cent for handling an extremely large and complicated estate. One winter night in about 1987, I met Johnny in the San Angelo Airport. He looked totally wore out. He had just come back from Montana trying to sell cows and the ranch for Gene's estate. He had fallen on the ice in Mile's City and broke his collarbone. Johnny, who had been a fighter pilot in WWII, a rodeo hand, and whatever else you might think of said, "You know this is the toughest job I've ever had. I know that someday Gene and I will meet again, and the first thing I'm gonna do is punch him the nose. After that we'll be best friends again."

Two significant negative events occurred in the early '70s as a result of changes in government policy. The government eliminated restrictions on grain exports to communist nations and subsidized the exports. The result was a rapid increase in grain price which reached its peak in the mid '70s, and also a shift in grain market premiums. Any export point like Corpus Christi or Houston, Texas became a premium market where previously the market for grain in the south Texas area was determined by the Panhandle price less freight.

The second major negative event was that as a result of meats boycotts and political problems, the government put a price freeze on beef at the retail level in June of 1973. Ever-optimistic feedyard owners assumed that when the price freeze was lifted, prices would go up rapidly but the opposite occurred. When the freeze was lifted, fat cattle prices dropped from 55-60¢ down to 35-40¢ while grain prices continued to increase. Gain costs that had previously been in the 30¢ range increased to over 50¢ per pound. The resulting cattle wreck in 1974 was catastrophic. Many feedlots and cattle feeders went bankrupt and the problems didn't stop there. Outside cattle numbers had risen to an all-time high and packer cow prices dropped to 10-15¢ a pound and feeder cattle prices to the low 20¢ range. At that time, as an individual, I was probably consulting for more cattle than anyone in the country and it was over a wide range of geography. I made a critical analytical mistake by assuming that the big drop in feedlot numbers would obviously result in a large increase in fat cattle prices. This was true for a short period in mid-1975, but inexpensive non-fed slaughter prices which included not just cows but also yearling and calves, overwhelmed the fat cattle market. Currently, the opposite is true and we now have a large shortage of "outside cattle" which tends to make the fat cattle market somewhat bullish even when feedlot numbers are relatively high.

In addition to the financial beating I took in 1974, the inability to sell fat cattle at any price left an emotional scar on my psyche. I had never experienced this before and promised myself unless it looked like a sure bet, I wouldn't own fat cattle again. Rather than speculate on feeding, I'd speculate on the futures. This was based on the assumption that if things went bad, at least I could unload my futures positions whereas I might not be able to sell fat cattle. My futures speculation rules were simple: 1) Never go long in months that you own cattle; 2) Never go short in months you don't own cattle; and 3) Never let your wife or girlfriend see your margin calls. Over the years, I've violated each of these rules—usually to my regret. However, even though my future's trading have had some rough times, I've seldom lost money on live cattle. Furthermore, in three years (1975, 1987, and 2003) I was fortunate to make significant profits feeding cattle. The 1987 profits occurred when land prices were low, and allowed me to buy a couple of ranches.

Around 1973 I began working with a new 25,000-head feedyard south of Eagle Pass, Texas. The owners were well-known in the area. In addition to the cattle business, they owned banks, and they also built a packing plant. It was a well-designed feedlot, but the owners were a bit on the wild side. Because of their proximity to the Mexican border and their Mexican connections, they fed a lot of Mexican cattle and had some pretty colorful border traders involved. As with most cattle people, 1974 was hard on them. I believe the feedlot closed in 1976.

Around 1975 I got an unexpected call from a friend working for the government.

She said, "Kenneth, I shouldn't be telling you this but your name is in a government file."

I replied, "What the hell is that all about?"

She informed me, "It relates to the time and people you supposedly associated with in the Eagle Pass area."

I was young and brash and mad as hell. I said, "Who can I call to find out more."

She gave me the phone number of a Director in Dallas. I called him, introduced myself.

"I am just a poor Aggie trying to make back the money I lost in the cattle business in 1974. Why is my name in your file?" I asked.

His reply was, "I have no idea but, I'll put you in touch with an agent in charge of that area. He's a good guy and will tell you what's going on."

I called him and he answered, "I have no idea what you're talking about and I don't recognize your name."

I then mentioned a couple of names of persons whom I had heard they were pressing charges against and he denied knowing them. Long story short, I decided they weren't telling me the truth, and if you really want to make me mad, lie to me about something important. Anyway, I decided there was too much drama in the Eagle Pass area for me, and I resigned my consulting position. I was sorry to do it because there were a lot of people involved whom I liked, but there were also rumors of other illegal problems in the area. One of the rumors revolved around the smuggling of guns from the U.S. into Mexico, and then shipping them to the Irish Republican Army.

Supposedly this was related to a large Las Vegas debt. I'm not certain the story was true, but it was a widespread scuttle butt in the area for several years. In the end some people went to jail on various charges. Moody Taylor went to jail for mishandling prepaid feed expenses. That's a charge a lot of cattle feeders could have been in trouble with at that time. Moody's lawyer told me it was obvious the law was going to get him for something.

If you're a country boy like me and always assumed that the law is your friend, this is confusing and makes you paranoid. I remembered thinking my phone might be tapped and told Andy and Brenda Binion Michaels. Brenda didn't make me feel any better. She said, "Hell, Kenneth, don't worry, they tap my phone all the time, and I don't do anything bad except smoke."

I thought I had quit Eagle Pass until one night in the late '70s my airplane was getting an annual inspection. I called a fellow in Amarillo who had said I could borrow his plane whenever I needed to. I called his house and fortunately didn't give my name—only asking if he was there. His wife replied, "No, he is in Eagle Pass with Dr. Eng looking at cattle." I asked her if he and Dr. Eng did that often and she answered, "Yes, at least once a month." I'm not sure there's a moral to this story, but it does prove you can get into trouble even when innocent.

Around 1970, Ed Cox, Jr. from Dallas Texas began making noises in the Texas cattle industry. Cox Sr. was a successful oilman in the Dallas area and the Southern Methodist University Business College is now named after him. Ed had a large ranch operation near Athens, Texas named Valley View Farms. Ed and I along with his right hand man, Rod Moss, became good friends and flew around the country in Ed's jet looking for investments and places to run cattle. That's how we happened to buy the ranch in California in 1971, but more about that later.

In the mid-seventies. after the '74 wreck, Mesa Petroleum was controlled by Boone Pickens. They owned several Panhandle feedyards and decided to get out of the feedlot business. Ed bought the 60,000-head Randall County feedyard for a song. In addition to the feedyard purchase, Ed also purchased a large silage and hay inventory at the feedyard. Their silage pits were huge and I was there when we opened the first pit—and what a surprise! Layered through the silage horizontally was sawdust. If I recall correctly, the sawdust layers were about two-feet thick and spread evenly through the silage pit. I remember Ed and I looking at this together and Ed asking, "Is this normal?" I replied, "Not until now."

Apparently the sawdust came from a sawmill on the Southside of Amarillo. The roughage which was supposed to be alfalfa also proved to be a problem. Long stacks of bales had alfalfa on the top but bales inside the stack were mostly straw and milo stubble. I'm sure Pickens didn't know this, but someone in the feedyard did. Ed asked for an adjustment, but I suspect they thought Ed had already made a great deal and he shouldn't ask for more. A lawsuit followed and Ed won. I've always wondered what was the real story behind the sawdust in the silage pit.

Forest Warren my oilman/cattleman friend and client from Houston bought a feedlot west of Amarillo on Interstate 40 and named it FSW. The F & S were for two other partners who later got into trouble and Forest brought them out. Forest sent his friend, Tommy Gleason, who had been a fat cattle buyer for Armour and a feeder cattle buyer to manage the feedlot. Tommy knew cattle but not feedlots. I told him first he had was to get control of the feed truck drivers because the feed delivery, feed distribution and damage to the bunks was getting out of hand. The first thing he did was pull one of the drivers out of the truck and beat on him. That didn't help. Besides the feedlot, they also had a grow yard east of Hereford on Highway 60. They milled their feed for the grow yard at the FSW yard and delivered it in feed trucks. One morning on his way to the grow yard, their feed truck driver tipped the truck over in the ditch by Highway 60. The highway patrol came by and gave the truck driver a DWI at 10 AM in the morning.

I told Tommy, "I think we've identified at least part of our feed truck driver problem."

When Tommy was buying fat cattle for Armour during the '74 wreck, he drove up to one of my clients in a new Oldsmobile.

"Tommy, that's a pretty fancy car for hard times." I said.

His reply was, "Whenever you see you're going to get into one of these really big wrecks, first thing you do is buy a new car because you'll be driving it a long time." Makes sense.

–K–

Heaven's Feedlot

K. S. Eng

When I finish life's final race
And my tired body is laid to rest
They say heaven will have a place
Where I can do what I do best

Feedlot life has been my fate
So the Lord should treat me well
They'll wave me through the Pearly Gate
Cause I've served my time in Hell

Calves will be preconditioned in Heaven
Health problems will disappear
"On time" truckers will unload them
Just stick a Temple tag in their ear

We won't need medicine or drugs
But company reps will come anyway
And leave some caps and coffee mugs
And take us fishing each Saturday

We will always have perfect feed
Thanks to Heaven's nutritionist
My old nutritionist I won't need
But he's still a welcome guest

Send no Holsteins each trader I'll tell
Bloats and bullers are such a pain
I know they'll still be hard to sell
Even in the Lord's domain

Heaven will have only friendly banks
Each week they will bring more money
Then buy our lunch and just say thanks
And their disposition will be sunny

With one old problem we still must deal
It's something the Lord requires
Heavens high standards are for real
So there's a shortage of packer buyers

Together In A Padded Cell

K. S. Eng

Some of my friends are reckless
But the friends I care for most
Are those who eat steak for breakfast
And skip the milk and toast

You wouldn't call my friends
normal or shy
They're gamblers who love
taking chances
They like it better when stakes are high
They gamble on cattle and ranches

The ladies we prefer are the lusty ones
Who stand face-to-face and toe-to-toe
The type that will stick to their guns
If you're willing, they're ready to go

Give me a lady who's independent
She doesn't need to stay at home
One who will help when I need it
But is able to do it alone

They don't care for lawyers
and contracts
Though you always know
where you stand
We've done a world of business
Sealed by a drink and a shake
of the hand

They're not what you'd call "high rollers"
But they like living on "the edge"
It may be safer in the valley
But it's more exciting on the ledge

They seldom "follow the beaten path"
They're often on the unpopular side
Sometimes loud, with a hearty laugh
But they're never short of pride

They're usually hunting another deal
Few are ever called lazy
Most are eccentric, but real
Some say, just a little bit crazy

Each of us know where we stand
And everything is ending well
We're all together hand in hand
Together, in a padded cell

-CHAPTER 10-

'70s Feedlot Consulting in Other Areas

One of my favorite consulting areas was the Northwest, in general, and more specifically Idaho. Joe Albertson who owned Albertson's food chain stores had a feedlot on an island in the Snake River near Nyssa Oregon named Gamble Island. His manager, Frank Hight, had a lease on the property. He had started feeding cattle before Joe bought it. Joe was probably the nicest rich man I've ever known, but obviously he could be tough or he wouldn't have been so successful. Frank Hight was a local boy who had gone through the Merchant Marines. He always loved water and landed on the island when he came back. He kept adding to the feedlot size until he ended up with 8,000 to 10,000 head. Joe said that was enough, and he asked Frank to get the cattle off the island because he wanted to make it a hunting preserve. He told Frank he could do whatever he wanted working for him in the future. Frank confessed that he just wanted to feed cattle, and to appease Frank, Joe built a 25,000 head feed lot on a hill near Nyssa, Oregon.

Two major events got me started on the Northwest consulting circuit—they were Urinary Calculi "water belly" and wheat feeding. Most of the feedstuffs in the Northwest tend to be high in phosphorus and, of course, the grain is low in calcium. At that time there were a lot of liquid supplements sold in the area, but they couldn't suspend calcium in them. As a result they had inverse calcium-phosphorus ratios and lots of water belly problems. I was able to solve the problem but not totally eliminate it.

There is a lot of wheat in the Northwest, and it tends to differ from wheat in other areas. Tom Hovendon, who was secretary of the Idaho Cattle Feeders Association, put on a great annual program. One year he had Ralph "Bull" Durham as main speaker. Ralph was promoting his all-concentrate feeding program. He made a round of several feedlots, and encouraged them to use an all-concentrate wheat-feeding program. The results were predictable and disastrous, and consequently, I started getting a lot of calls and new clients. I also had some experience with potato feeding and of course that is big in the Northwest especially with Simplot. My Idaho clients included Albertsons (Frank Hight), Rich Hormachea, Gary Chipman, Simplot (John Basabe), Richard Blinco and Ralph Harding. I learned that the Idaho folks not only knew how to feed cattle, they also knew how to play and party. When in eastern Idaho, we often stayed at the Holiday Inn in Twin Falls. During that time, Evil Knievel was preparing for his motorcycle jump across the Snake River near Twin Falls. Knievel would sometimes join our party, and obviously, he was a daredevil, but in that group he was the quiet one.

You could write a book on any of these individuals because each was colorful in his own way. Joe Albertson finally announced to Frank that he was tired of feeding cattle and was going to sell the new feedlot. He would find another job for Frank doing whatever he wanted to do. However, Frank still wanted to be involved in the cattle business, and he had enough Albertson stock to make himself comfortable. He fed cattle around Caldwell for a few years and began to concentrate on grazing yearlings in Nevada. Even when he quit feeding cattle, I still stayed with Frank and Doris when I was in that area—they were great people. Frank had some good grazing leases in the Ruby Valley area of Nevada, and over the years he made good money on yearlings.

At one time, Nevada did not have a speed limit. After they instituted one, Frank was legendary for his speeding tickets. I was talking to Frank one day, and he said, "I might be in trouble. I got a ticket for going a 110 mph near Elko, and I've got to appear in court." I advised, "Frank, you need a cruise control for your car," and he replied, "I've got one but it doesn't work. I had it set on a 100 mph and damned if they didn't ticket me for 110 mph."

Albertson's sold the Nyssa Oregon feed yard to Victoria Cattle Company and they brought a man in from California to manage it. They asked if I would stay as a nutrition consultant. The new manager had a lot of feedlot experience in California, but did not get along well in the Northwest. The feedyard was on a steep hill with very little winter protection and occasionally it would really get cold and windy. In October the manager and I were driving around the feedyard and I noted they brought in almost 10,000 head of Brahma cross and Mexican cattle.

I said, "These cattle could really be a wreck if you have a bad winter."

His reply was, "You balance the damn rations and I'll take care of the *#@$%^ &%$@** cattle."

When I went back to the office, I gave him some ration changes and turned in my resignation.

"You don't need to be so sensitive," he told me.

I said, "I'm not, but I've seen enough wrecks the past few years, and I don't want to be here for this one."

They shipped most of the cattle back to Bakersfield four or five months later because they were sore footed, foundered, and the like. Not long after this time, Victoria filed bankruptcy.

John Basabe was a rough, tough Basque who worked and played hard and J.R. Simplot loved him. Simplot had several feedyards in Idaho, including his first one near Caldwell that was run by Bill Richardson. Each of the feedyards fed a lot of potato waste. Richardson and Tom Hovendon were big buddies and liked to party at the Saratoga in Caldwell. One night some hippie-looking kids from California came into the bar, and Bill asked them what the hell they were doing. They admitted that they had been duck hunting but had only shot a few teal. Bill said, "Come with me, I can get you more geese than that with my pickup." He drove out to a pasture full of geese below the feedyard. He made a figure eight through them and I believe they gathered up sixteen dead geese. Anyway, Bill sent the hunters home happy.

I learned that if you made a suggestion to Basabe that he liked, it would be implemented on a large scale immediately. He called me one night saying, "You've got to get up here, there's something wrong with our potato pit." When I arrived at their large feedyard below Grandview, there was indeed something strange in the potato pit. The potato plant had changed the process for peeling the potatoes from steam to a lye peel that raised the pH of the potato peel waste to an incredibly high level. Since Simplot produced phosphoric acid fertilizer, I figured out how much phosphorous acid to add to the peel waste to get the pH down to supply supplemental phosphorus. I meant this to be mixed at the potato plant in Caldwell, but John decided to add it directly into the pit at Grandview. He sent an employee into the pit on a tractor pulling a weed sprayer full of phosphorus acid. Predictably, the tractor sank in the middle of the pit. We had a photograph taken by Baxter Black of the driver standing on top of the tractor hood just before it went out of sight. A rope was thrown to him, and he was pulled him of the pit. I don't know what happened to the tractor and sprayer.

Baxter had just joined the Simplot staff as feedlot veterinarian, and he was doing his best to please John and get along with me as well. He and his first wife lived in a house up the hill from the feedyard so he could never get far from the job.

One Sunday afternoon John picked me up at the airstrip, and we went to the feedyard.

"Oh, damn," John said, "I was supposed to go see Baxter this morning and I forgot." He called Baxter and apologized.

Baxter answered, "That's okay, because it got my wife to take down the Christmas tree." This was in April.

Tom Hovendon, Secretary of Idaho Cattle Feeders for several years was also a writer for Champ Gross at Calf News. He loved the feedlot industry. He was one of the best writers I've ever known, and he had a great sense of humor. Tom who was originally from Nevada related a great Nevada story about three fellows who decided to go to Winnemucca to enjoy the entertainment. Later that weekend they were fishing with a fourth man who was an MD when one of the fellows mentioned he was getting some bad pains in his "private parts" area. The doctor asked if they had visited the Red Light district in Winnemucca. When they confessed, "Yes." He advised them to come in for penicillin shots. Also, if you need to, bring your wives." It was probably good advice except for one thing, it turns out the fellow with the pain had kidney stones.

At one time in the late '70s I believe I had approximately thirty clients in eighteen states. If this is not the definition of insanity, at the very least it's the game plan of a fool. Obviously, I wasn't spending the time I should have with family and people who care.

As I said before, I'm going to spend more time in this book recounting humorous events because I truly believe that "laugh and the world laughs with you, but weep and you weep alone." Perhaps my funniest conversation occurred at a feedyard located near Great Bend, Kansas. Tom Taylor owned and managed the feedlot.

When I arrived one day he apprised me, "We've got a customer from northeast Iowa, and he's got a couple of pens of fat cattle he won't sell. I'm going to arrange for him to ride with us and we'll stop in front of his fattest pen. You tell him why he needs to sell them."

We stopped in front of a pen of his overfed Herefords, and I pointed out that feed prices were then at an all-time high and gain costs would be astronomical if these cattle weren't marketed soon."

His reply was, "Young fellow, let me tell you something. Back in the early '50s I sent a trainload of cattle to the Chicago stockyards, and they screwed me around. So, I took them home and fed them another year, and you know what—cattle don't eat much that last year."

Although he didn't mean it to be, this was both a funny and an astute observation because once cattle reach their physiological end point of finish, their consumption drops dramatically. I had a friend that fed a small group of Angus steers on a finish ration for four years. He said the consumption went down to a maintenance level after two years on feed and they didn't gain or lose a pound the final year.

One of my favorite small and out of the way clients was Western Slope near Grand Junction Colorado. Bill Larimore was the manager and part owner of the feedyard. The other owners were the Boice family from Arizona. Pancho Boice was the de facto head of the group. He was killed in a plane crash on the way to the feedyard. They grew and finished cattle at the feedyard, and once a year they would receive about 4,000 Arizona Hereford calves to finish (these were the old-style Herefords). At that time there were so many packing plants they could shop for a new buyer each year. However, in '74 the only way they could sell them was "in the beef" at the Excel plant in eastern Colorado. This is the first time we got to look at these cattle on the rail and the results weren't pretty. The majority of the carcass quality grade was no roll or good, the cutability was 4's and 5's and there were numerous docks for light carcass weights.

They bought a large amount of silage each fall from local producers, and one year they asked me to develop a percentage formula for premiums and discounts based on silage dry matter. I asked them what they wanted to use for an average dry matter and they said 30%. They were paying $14 per ton delivered to the pit. I said a 1% percent variation in dry matter is approximately a 3% premium or discount. I suggested using no premium or discount for 29% to 31% dry matter silage and then applying premium and discounts at the rate of 3% per dry matter percentage point variation. In other words a 35% dry matter product would have a five times three or 15% premium, and a 25% dry matter product would have a five times three or 15% discount. They liked the idea and that's the last I heard about it until about 7:00 o'clock one evening. They called from their office and said they were celebrating the final silage delivery.

Bill informed me, "Your damn formula doesn't work."

"Why?" I asked.

He said, "Because the last guy's replanted silage was only 25% dry matter. With a $3 per percentage point deduction that's $15 per ton. We're only paying him $14 base price so he ends up owing us a $1 per ton for the silage he delivers."

My heart sunk, but I dug out the file, looked at the formula and saw that it read "*percent*" not "*dollars*." They had reprinted their own contract and substituted the dollars for percent. Fortunately, most of their silage had been very consistent around 29% to 31% dry matter so the premiums and discounts didn't mean anything until the last field. Anyway, they went back to their bourbon and the fellow who thought he was going to own them a $1 per ton for delivering his silage was pacified.

–K–

Pretending

K. S. Eng

As I watch the market fall apart
I go on and keep pretending
Told the wife it'll come back
This can't be the final ending

The cows are grazing rocks and dirt
It's been a drought without an ending
And the packers control the markets
They said, "Trust me" but were
pretending

My ex-partner owes us money
He promised he'd soon would send
But "Hell" he's just as broke as me
All he can do is just pretend

Told the wife money was coming
We've a land sale that is pending
But creditors had tied it up
The wife said I was pretending

If my banker will go one more time
We could be on the road to mending
I'm broke, but maybe my credit's good
Lord, I hope I'm not pretending

I asked them to extend our loan
I hoped they were still lending
They said, "You're way past broke
It's time you stopped pretending"

The wife and kids are moving to town
They're tried of the trouble and pain
I hoped they'd love the ranch like me
But it's hard with no money or rain

Guess I'm part of a "dying breed"
And as they say at the final ending
It's " ashes to ashes and dust to dust"
There's no reason need to go on
pretending

When they close the coffin lid
And the casket is descending
The wife may say, "Check it again
Let's make sure he's not pretending"

Waiting for the Market to go Up

K. S. Eng

We had plans for a big anniversary bash
But as she refilled my coffee cup
She said prices are low and
we're short of cash
Let's wait for the market to go up

We were going to a restaurant
fancy and new
When our work was all done last fall
But the lady said staying
home would do
Cause the check for the calves
was so small

There's meat in the freezer
I'll fix tonight
I'll get the silver and set the table for two
We can enjoy a steak by candlelight
It will be cozy just me and you

You learn living the country life
To cherish true friends
cause they are few
And through the down markets,
trouble and strife
My truest true friend was you

I guess you've been gone nearly a year
And though the market has
dealt me some luck
I sit here alone sipping my beer
Caring little the markets gone up

Now I can afford restaurants
fancy and new
With the up market and the
money I make
But I'll just set the table for two
Relive old memories and cook a steak.

-CHAPTER 11-

Clayton and Union County, New Mexico—Feedlots, Grass, Bankers and the James Family

In 1970 Clayton Cattle Feeders asked if I would do their consulting. That began a long association with the people and livestock industry in Union County and the surrounding areas of New Mexico. I was contacted because they bought a Cabot Popper to process their milo and I was working with the original feedlot (Veribest Cattle Feeders) that had installed one. Cabot recommended me as a consultant. Clayton is the only town of significance (population 2,500) in Union County and was also home for five feedyards plus the center of some of the best short grass grazing in the United States. Over the next thirty to forty years I became acquainted with many great people who were also colorful and tough. I fed and grazed over a 100,000 head of personal cattle in the area and I also ended up with partial ownership of two feedlots.

Union County has a unique high desert climate with 10–15 inches of annual rainfall and low humidity because it is at 5,000+ feet altitude. On average, the climate is great but there are exceptions and extremes. I've seen severe blizzards as early as late-October and as late as mid- May. We had a Halloween blizzard one year that killed approximately 100 of our yearlings on excellent pasture. The storm then went straight across the Texas and Oklahoma Panhandle. The length of the storm was approximately 120 miles and the width only about five miles. I had a client (Forrest Warren) with pasture cattle north of Stratford, Texas who was scheduled to ship the first of November. There were 504 in the group he was shipping, and when the blizzard was over, only four were left alive. No storm insurance. What made our deal tough is that we had winter insurance, but it did not start until the First of November—the day after the blizzard. This storm occurred in the mid- to late-'70s while we were still recovering from the wreck of '74. We didn't need the added grief. Another severe blizzard occurred at Easter in the mid-'90s. I had several thousand light yearlings with Benny Gilbert in northern Union County. Caroline and I were in California and Benny called me about four in the morning California time. I asked him, "How's the storm?" His reply was, "Doc, it's not a pretty picture." He said he was lost and thought he was right on the edge of a canyon in his pickup and didn't dare move. He sat in his pickup for several more hours until his hired man came and sure enough, his pickup was about five feet from canyon edge. We lost approximately 150 calves in that storm—but more about that later.

The ownership of Clayton Cattle Feeders was initially twenty stockholders who each had equal shares. The group was made up of local ranchers and businessmen who were pretty smart and damn tough. After the wreck of '74, several wanted out and three of the stockholders, Charlie James, the Manager; Clyde Sowers, rancher and owner of the *Union County Leader* newspaper and Bill Shaw from San Angelo, wanted me to buy any stock offered by the other stockholders. I didn't have much money, but it didn't take much and I wanted to own pens with a couple thousand head capacity for experimental purposes. In the end, it all worked out well, but there were some bumps along the way. At the first stockholder's meeting I attended, they spent the entire time in heated arguments. I vowed never to go to another stockholders meeting, and I gave Charlie James my voting proxy. It was unusual that Clyde Sowers came to the meeting with a briefcase. When we left I mentioned to Clyde that he had never opened his briefcase. He replied, "I didn't have to, but I had my .45 in there in case things got out of hand." The four of us ended up owning the feedlot plus a wheat-native pasture ranch nearby. When Clyde died, his wife, Betty, inherited his shares and we decided it would be good to get key man life insurance policies. When the insurance agent made her presentation and we accepted it, we ran into a glitch because Betty wouldn't give her age. She said it was none of our damn business—so we estimated her age and got by.

Feedlot luncheon

Charlie was a good manager and the feedlot stayed jammed full of cattle with a customer waiting list for many years. In the mid- to late-1980s, several people wanted to buy the feedlot. I thought the price they were offering was fair, so I told Bill and Charlie I thought we should sell. They asked if I would be mad if they bought my portion and kept theirs. I said, "Not in the least." I consequently ended up with a nice profit that helped buy a southwest New Mexico Ranch. Later Bill and Charlie got crossways and Bill brought Charlie out. I was sad that what once was a great partnership and friendship ended in total disarray.

Clyde and Betty Sowers loved to throw parties. At one of their Christmas parties, Betty ran out of Frangelico and sent my pilot to the local liquor stores to buy a bottle. They didn't have any in Clayton, and I found out years later that Betty then sent my pilot and plane to Amarillo for a bottle of Frangelico. I never paid much attention to the hours or expenses with an airplane, but that was an expensive bottle of liquor.

Betty died of cancer several years later. What made this especially traumatic was that her son who was in his late '40s and was running the newspaper died suddenly three days before Betty died. No one wanted to tell Betty the bad news as she was in and out of consciousness. The lady taking care of Betty said she kept saying, "I hear my baby crying and I need to go." That still sends shivers down my spine.

Another great friendship and business relationship was with Rex and Vi Reeves who had the First National Bank in Clayton. Rex was one of a kind, and if he liked you, you couldn't have a better friend or banker. If he didn't, you were S-O-L.

In late 1974 Rex called me into the bank and said, "Doc, you're almost out of cattle."

I answered him, "I know, I'm broke."

He replied, "You find the cattle, and I'll find the money."

I then proceeded to sign twenty blank notes and started buying calves in late 1974 when they were dirt cheap. The first group I bought was purchased by Rod Moss who was buying for Ed Cox in East Texas. I happened to stop by the sale barn and I asked Rod Moss to put together a couple of loads of heifers for me. They cost seventeen cents per pound and delivered to Clayton weighing approximately 450 pounds. They should have weighed 700 pounds, but they were thin. Along with other cattle I bought late in '74 and sold in June of '75 when there was a brief uptick in the market made about $150 per head. Thanks to Rex, I recouped all my losses. Perhaps the biggest personal casualty was that the experience traumatized my wife, Pat. Her family had gone through financial problems in Nebraska in the early '50s. I don't know why, but financial risks and losses never worried me because I always felt like I could somehow make it back. Perhaps it's because I've never cared much about money as such. After we divorced, Pat no doubt slept better.

Rex and Vi had some real hard times with the bank partially because Rex would finance almost anyone for an extra one to two percent interest if he liked them. As a result he had a lot of cattle customers who were sort of a rogue's gallery. Whenever cattle times got tough, it was tough for the bank. Rex and Vi separated in the late '80s during banking bad times and Rex later committed suicide. His funeral was on April 15, 1988 (tax day). Vi and her son, Craig, ended up taking over the bank and have managed to expand. I believe they now own six banks, and I'm really proud of our friendship and their success.

I went into Clayton at least once or twice a month for over thirty years, and I ended up knowing lots of people. The old Eklund Bar and Hotel is where Black Jack Ketchum was hung in 1901. It's a landmark and good watering hole. One night we decided to have our feedyard board meeting at the Eklund Bar during a severe blizzard. The fellow who owned the bar said he wanted to go home and we were the only customers. He gave us the key and told us to serve ourselves and let ourselves out. We tried to drink at least one shot from every bottle at the bar and by sun up we were ready for a blizzard of any size.

What Went Wrong?

Black Jack attempted his final train robbery by himself on August 16, 1899.. He was surprised in this attempt by conductor Frank Harrington, whose shotgun blast demolished Ketchum's right arm. The next morning he was arrested and taken to a hospital in Trinidad, CO. While he was being held at the New Mexico Territorial Prison in Santa Fe, his wounded arm became so infected that it had to be amputated.

Black Jack's hanging was Clayton's first (and last). No one in the area had any experience with hangings. New Mexico Governor M. A. Otero sent Lewis C Fort to Clayton to oversee the arrangements. That didn't help.

A drop of five feet would have been sufficient…Ketchum was dropped seven. It's possible that the rope was too thin. It was tested the day before with a 200 lb. sandbag, which was left on the rope all night, stretching it and removing its elasticity.

It's also possible that the noose had been soaped to ensure that it would slide well…perhaps too well.

Tom Ketchum was the only man on America's western frontier legally executed for a crime other than first-degree murder; and his execution was the most bungled affair in the annals of frontier justice.

Black Jack Sites to visit in Clayton: Directions to the 1901 hanging site of Black Jack Ketchum: From the county's only traffic signal go west, on main street, cross the railroad track; once across track, veer left immediately onto Court St. Stop sign on Hwy 402. Beyond stop sign on your left is the Union County Courthouse. On the west side of the complex is the small, square brick building of the sheriff's office. The scaffolding for Black Jack's hanging was erected against the north wall. Notice the plaque on the wall commemorating this site (placed there in 2001).

Directions to Black Jack Ketchum's Grave: Begin at the traffic signal in downtown Clayton. Go east along Main St. (Highway 56-64 toward Boise City, OK) Follow signs toward Port of Entry, Airport, Golf Course and turn right on West Avenue. Keep following signs – Airport, National Guard Armory– and turn left on Water Street. Pass the trailer park and descend hill to the stop sign. Turn right at Stop Sign (do not go through the yellow gate.) You are now driving along the west side of the golf course. The main entrance of the cemetery is on your right, marked by a brick gateway and a flagpole. ENTER: Go straight west beyond the flagpole and along the tree row. About halfway down is Black Jack's Grave, under the trees, and in the middle of the road.

Additional information may be viewed at the Clayton Chamber of Commerce Visitor Center, The Herzstein Memorial Museum and the Hotel Eklund.

ENJOY YOUR VISIT TO CLAYTON, N.M.

Thomas Edward Ketchum
Train Robber and Murderer

Born on Halloween, 1863 in San Saba county, Texas. Tom and his brother Sam's outlaw career spanned the years 1896-1899

Tom was hanged in Clayton, New Mexico on April 26, 1901 in a bungled execution in which he was decapitated.

The hanging site (the north wall of the sheriff's office at the courthouse) may be visited today, along with Black Jack's grave at the local cemetery east of town.

Tom Ketchum's Hanging

The man adjusting the noose is W.H. Reno. Colorado & Southern Railroad detective, who had pursued Black Jack for several years. The man on the right is Tom Gray, a local saloon owner, who was also deputy county sheriff.

The sheriff of Union County at that time was Salome Garcia,, a religious man who was opposed to capital punishment.

The event was a circus. People bought tickets to view the hanging from inside an enclosure, and kids bought souvenirs of a wooden stick man, hanging from a string on the end of a wooden stick.

The Decapitation

On the left, barely visible in the very dark area is visiting sheriff John McCandlass of Dalhart, Texas.
When Ketchum fell through the trap door his head was jerked off, his body landed on its feet, then fell over and began convulsing violently.
McCandlass grabbed the body by the shoulders and helped to minimize the convulsions until they stopped.

Leaning over the body is Dick Franz, who sprung the trap door for his friend, sheriff Garcia.

Next, half standing, is jailer Ramundo Arguello.

On the right is Dr. Slack, fulfilling his duty as county coroner by taking the headless body's pulse. It reportedly took five minutes for Tom Ketchum's heart to stop beating.

The Aftermath

Left: Sheriff John McCandlass of Dalhart, Texas

Right: Dick Franz, friend of Union County Sheriff Salome Garcia. Franz cut the rope which sprung the trap door, hanging and decapitating Tom "Black Jack" Ketchum. It is reported that he enjoyed his celebrity status as the man who actually set the hanging in motion. Notice the black blood-soaked hood in the foreground containing Black Jack's head.

After the head was sewn back onto the body, Tom Ketchum was buried at Clayton's Boot Hill. In the 1930's, due to housing development, his body was moved to the new cemetery in Clayton, where it remains today.

The legislation which decreed death for train robbery was later repealed.

For further reading, the definitive book on the Ketchum Gang is "The Desperate Outlaws" by Jeffrey Burton

Clayton, although small, was sometimes the center of some extreme violence. I think there were three or four killings near there in a ten-year period several years ago. During that time, my pilot, Beadle, and I had separate rooms at the local hotel. One day I told my pilot I would call his room in the morning when I was ready to leave. No one answered in his room when I phoned. This went on for a couple of hours, and I finally asked the hotel to check his room. They found it hadn't been slept in. We mounted an all-out search for him because at that time a couple of the murders in the area remained unsolved. We worried something bad had happened to Beadle. As it turns out, the motel had checked him into the wrong room and he was sitting watching television and wondering why I hadn't called. We found this out after we had checked the local hospital, jail, all of the restaurants and bars and several of the country roads.

Gene Atchley was raised west of Clayton, New Mexico during the dust bowl and Depression days. After World War II he spent his entire life working in the cattle industry in New Mexico and Colorado. Before I hired him to manage the feedyard we built with Rex Reeves assistance, he had worked for Whitey and Gary Wisebart for several years. He and Gary Wisebart, who now lives near Santa Fe, can tell some great stories.

Gene and his wife, Verleen, were close personal friends of Caroline and I. Verleen died not long after Caroline drowned. I love to hear Gene talk about the ranch days when they were still shipping calves by rail from around Mt. Dora, New Mexico to Kansas City. Gene said they would split off the calves from the cows and then run them hard several miles so they would wear down and not turn back to their mothers. They would then move them into the pens by the railroad track. Usually the train would be late. The calves were then shipped to Kansas City by rail, put on trucks and typically ended up in Iowa and Illinois feedyards. In spite of that rough handling, Gene said no one ever complained about health problems. This tells you something about the changes in our calves over the past several years.

Harold Parker and I fed quite a few calves at Gene's and Charlie's feedyards. Harold would buy them in Mississippi—and with Harold involved there was often some drama and surprises. One time Harold had put a load of calves onto a truck near Hattiesburg, Mississippi to send to Clayton one weekend. Harold and I were to partner on the calves, and they were to be sent to Clayton Cattle Feeders only two miles down the road from Gene's feedlot. The trucker ran into a bad blizzard and probably other problems that are better to not discuss. He showed up at the feedyard late Sunday night—two days late. He turned into Gene's feedlot which was the wrong one, but they had the receiving gates set for some other cattle they were expecting. The trucker got them unloaded by himself. There were several dead cattle on the truck and he dragged them off and left a note tacked to the feedlot office door saying, "Here's your damn calves." Monday morning Gene called me and we finally got the situation figured out. I then called Harold.

Harold explained, "Well, this is a little bigger problem than you might think. We were supposed to pay for these cattle based on their off truck weight and the cattle were loaded in three separate groups and were to be weighed separately."

I informed Harold, "That can't happen now, and there's the additional problem of dividing up the dead calves."

Harold said, "There's another problem, we can't find the truck driver and I want to fire his ass."

Harold called back several hours later to say that they had found the truck someplace in Mississippi with a note on it saying to "Call Harold Parker." The trucker knew he was in trouble, and he just walked off. We tried to put everything back together, but there were more questions than answers.

Finally Harold called again to inquire, "Doc, what do you really think."

I told him, "Harold, I think it's a wreck."

Harold replied, "Well, I know that, but do you think it's something money can fix?" I added that to my list of favorite Parkerisms.

I said, "Money can fix it easier than we can figure it out."

About six years before Rex died, he repossessed a feedyard that he had sold to several traders that predictably didn't work out. He gave me 49% of this feedyard, requesting I find a manager and assistant manager then splitting my half with them. We renamed it New Mexico Cattle Feeders and it was quite successful with Gene Atchley, Manager and Freddie Polling, Assistant Manager.

Three or four years later Rex called me saying, "I need all the shares back to pledge on some loans. What do you, Gene and Freddie want for your shares?"

I suggested to Gene and Freddie, "Let's make him an offer he can live with and take what we can get as he'll end up with it anyway."

It worked out fine for Gene, Freddie and I. However, Rex still went broke and we were at a loss to understand why because the feedlot was making good money. We never knew where Rex lost his money, but he liked to partner with some pretty high-risk characters. Even still, he didn't have the usual bad habits such as futures trading, gambling, girls, and the like. Regardless, I miss him and he was the best banker I ever had. Were it not for the blank notes he had me sign in late 1974, I couldn't have continued to own cattle. God Bless your soul, Rex!

The banking business was tough in Texas and New Mexico in the late 1980s and they were threatening to close Rex's bank. Part of it had to be sold to outside investors, and Rex and Vi split up. Rex also lost his position on the board and that depressed him. Rex often carried a revolver and both Gene Atchley and I were sad to hear he committed suicide, but glad he didn't take someone with him. Because of bad loans to a lot of high-risk friends Rex was in a lot of trouble, and I'm sure he felt that many people had let him down. An interesting tidbit is that one of the things that helped the bank survive was the fact that Rex had taken out a $600,000 life insurance policy on a border trader to cover his overdue loans. When the trader was killed when his private plane crashed, the insurance money helped save the bank.

The event that put the Eklund Hotel on postcards was the hanging of Black Jack Ketchum—a train robber and cattle rustler. He was hung in 1901 near the Eklund Hotel. Because they didn't adjust the noose properly, his head was jerked off when they pulled the trap door. Postcards chronicling the event are found in many Western bookstores. It is said that Black Jack's final words to the crowd were, "I'll see you in hell for breakfast."

No stories on Clayton, Clayton Cattle Feeders or Charlie James and family would be complete without mentioning Charmayne James. Charmayne first won the barrel racing championship at the Las Vegas National Finals in 1984 when she was fourteen years old. She and her horse, Scamper, continued an unbroken winning streak at the National Finals for the next ten years. She then retired Scamper continuing to compete for approximately nine additional years winning an additional championship on the different horse. This record will likely never be broken and Charmayne and her mother, Gloria, did it the hard way. They probably traveled 80,000 miles or more per year with a pickup and horse trailer for almost twenty years and a total of more than 1.5 million miles.

The story of Charmayne and her horse, Scamper, and the James family had a fairy tale beginning. Charlie, who has a great eye for horses, bought Scamper from a feedlot cowboy for $1,100. The horse was four or five years old and was too much for the cowboy to handle. Charmayne began riding Scamper shortly thereafter. She had a set of barrels in their backyard at the feedyard. She and Scamper would practice every day. I recall a funny incident when I was talking to her one time when she was practicing. She had a pet goat.

I asked her, "What do you plan to do with that goat?"

She replied, “Raise little goats.”

I said, “But you only have one.”

And she replied, “That’s okay, it’s a girl.”

Scamper died on July 4, 2012 at the age of thirty-five, but not before they cloned him. They are collecting semen from the clone so we haven’t seen the last of Scamper. Charmayne continues to conduct barrel racing clinics throughout the world, and she and her family along with Charlie and Gloria live on a ranch near Boerne, Texas.

Some of the best advice I ever gave clients was in 2007 when I encouraged Roger Schmidt and Gene Atchley to sell their Clayton feedlots. Feedlot demand was high and they got great prices. Shortly thereafter the market collapsed and while I miss them as clients it was a good business move.

–K–

Bank Inspector

K. S. Eng

My title is cattle inspector
I work for the big city bank
They sent me to the Diablo Ranch
Whose owner was said to be rank

I phoned and he said "come on out
I'll meet you at the gate
You need to be here at sun up
I'll be waiting so don't be late"

"Cows are easy to see when it's early
Climb in the pick-up and let's go"
Then he took a sip of coffee
And jammed it in 4 wheel low

We started down a dirt road
And hit every bump and rut
By time we reached the next gate
I had blisters on my butt

Next we crossed a creek
Doing 40 to keep from sinking
He said, "be sure to count the pairs
Standing in the creek a drinking"

I opened another gate
At the top of a 40 degree grade
Almost stepped on a rattlesnake
That was headed for the shade

We started down an arroyo
I bit my tongue and my lip
Cause we hit it doing 50
He called it a little dip

Baling wire flew from the pick-up bed
Along with everything not tied down
We lost beer cans and a cooler
And a Catahoula hound

We were doing 60 on an S curve
He reached in his pocket for a "chew"
I mentioned the front tire was bald
He said "Don't worry the spare is new"
As he pointed to the South he said
"The rest of the cows are behind
that bluff
You had better cinch up your seat belt
That country gets a little rough"

I replied, "That wouldn't be necessary
I've seen enough of your herd
You know what they say
about cowmen
You can always take their word"

When we got back to the front gate
I was bruised and had the shakes
As the pick-up slide into the bank car
He mumbled, "Gotta fix the brakes"

I staggered back to the bank
And asked them to call me a doctor
Told my boss I'd finish the count
When he got me a helicopter

Country Banker

K. S. Eng

I miss my old country banker
Cattle business isn't much fun of late
Old times weren't always perfect
But he and I shared the same fate

Like back in the wreck of 74
I was out of money and collateral
He said "I'll find you the money
You need to keep buying cattle"

Told me to "Buy when the price is low
And when you think you need more
Just call me whenever they figure
Leave some signed notes in my drawer"

He wasn't much for Financial
Statements
Said they were mostly "fluff
and blue sky"

He would rather shake a man's hand
While looking him straight in the eye

You knew that he would stay with you
No matter how rough the ride
Lot's of changes have taken place
Since he crossed the great divide

I finally have a fancy statement
Prepared by my accountant & my wife
Guess you have to have them
It's just a fact of life

Now I shake hands at our loan meetings
Nod to my wife and thank her
Then I head for a quiet bar
To toast the memory of my
country banker

-CHAPTER 12-

Confinement Barns and Calf-fed Holsteins

I've been blessed in my career with good timing allowing me to be present when new breakthrough events were occurring. Two of these events began in the late '60s and early '70s—calf-fed Holstein programs and confinement slotted floors. The early pioneers in this business were a father and son team from California. Their background was not in the cattle business but in aviation. Dale Smith was an early barnstorming pilot, and his son, Dale Jr. was a test pilot for Howard Hughes. Dale Jr. was a bit wild and a good friend. In the late '60s they built the first slotted feedlot south of Los Angeles. They found their source of calves in the southern California dairies. Most of the calves cost $5-$10 per head and you could select only the best. The calves were placed on feed when they reached approximately 200 pounds and they sold at a finish weight of 900-1000 pounds. At that time southern California packers discriminated against heavier carcasses and did not particularly want choice carcasses. The calves were finished on a high concentrate diet and the conversions were great.

At that time Californians were the innovators in the commercial feedlot industry because they had an abundant supply of Holstein calves, a mild climate, a demand for lean carcasses and an abundance of available byproducts such as beet pulp, citrus pulp, almond hulls, molasses, and fat among other choices. The Arizona cattle feeding industry developed shortly after the California cattle feeding industry since they had a similar climate to southern California and even though they didn't have the variety of by-products available, they produced a lot of milo, barley, wheat, alfalfa, cottonseed products and the like. Gene Erwin and I were the first to feed whole cottonseed to cattle in commercial feedlots in Arizona. We brought the cottonseed at a $10 discount to grain. Later dairies started buying whole cottonseed driving the price up substantially.

Beginning in the early '70s there was a stampede to build slotted floor confinement facilities throughout the United States. Designs included slotted floor, teardrop floor and flume floor. None of these were wildly successful, and if you travel the back roads, you will still find remnants of old confinement facilities that were closed after a few years.

One of my favorite slotted floor stories involved particular clients of mine in Sikeston, Missouri—a town down in the boot heel of Missouri. They called asking if I would consult with them on a new feedyard they were building. I checked the average rainfall for the area. I believe it was somewhere between 40 and 50 inches, and I told them I thought it was too wet in that area for a feedlot facility. They said not to worry because they planned to build on a sandy ridge that has never had mud on it since the Civil War. My reply was, "It probably hasn't had a feedlot on it either."

Unfortunately I was correct and mud became a serious problem within a year. They built a confinement facility that was moderately successful. The best news is that we became good friends. Glen Barks was part owner and manager, and I always looked forward to my Sikeston trip even though it was probably my smallest client. I always felt that the lasting relationships you build with consulting clients are more valuable than financial considerations.

The slotted floor design of the Smith Feedlot consisted of concrete pens approximately twenty feet deep with floors made up of approximately two feet of solid floor in the back and front and sixteen feet six- to eight-inch concrete slabs with a two-inch opening between them.

The combination of calf-fed Holsteins and slotted floors worked great, however, it should also be pointed out that most calf-fed Holsteins do well in Southern California and Arizona. Because of their thin hide and low back fat, Holsteins withstand the heat well, but also, for the same reasons, they do not do well in cold weather unless they are well protected. Dale, Jr., who was never short of new ideas and desire, decided to move their program to Northwest Iowa where corn was cheaper. It's good to remember that if you're in the calf-fed Holstein business you're also in the feed business because they consume a total of two to three tons each of feed during the normal feed cycle. Smith's new feedlot near Cherokee, Iowa had approximately eight- to ten-thousand head. Their local partner was Dr. Bob Bryant, a veterinarian in that area. One other unique part of their slotted floor design was that they had scrapers under the floors that pulled the manure into trenches then fed into lagoons. The Iowa operation was fairly successful, but they soon discovered that in cold weather, the scrapers froze and then manure accumulated on top of the slots and froze. Plus, the concrete was very cold. The facility was completed just in time for the 1974 market wreck resulting in the big increase in corn prices in 1975 and 1976. A Holstein calf suddenly became worthless because of high corn prices and the low fat cattle market. Stories abound of people bringing small calves to the auction and putting up a sign for free calves! When they came back later they found that instead of getting rid of their calves, people had added more calves to the group.

These were tough times for everyone—especially Holstein calf feeders. One humorous event involved a calf operation Smith had bought that was approximately three hundred miles northeast of Cherokee, Iowa. They equipped a barn to raise approximately twelve hundred day-old Holstein calves. The person who was supposed to be operating the facility had not been heard from for several weeks. They decided to take a personal look at the operation and we flew up in my plane. What we found was mind-boggling. There were no people around and fortunately the calf crates were empty. However, several other interesting developments had taken place in their absence. A creek had been diverted so that it ran through the dining and living room of the main house. There were iron grates over the running stream that then went out into a concrete pond behind the house. The concrete pond was filled with large rainbow trout and next to it was a trout hatchery. I was told that the man in charge was selling limited partnerships in a trout breeding program. There was also a new road leading to the top of a mountain behind the calf barn. On the other side of that hill they had started building a restaurant and a ski slope. Needless to say economic circumstances precluded the completion of these projects and I've always wondered what happen to the facility.

Holsteins are still the most popular feedlot animal in Southern California and Arizona for obvious reasons. There are large dairy herds close to this area therefore source of Holstein calves and Holsteins tolerate hot weather very well. However, large dairies have moved to other areas like Idaho, Texas, New Mexico, Illinois, Indiana, etc. and so has calf fed Holstein feeding.

Holsteins differ from beef cattle, but there are several reasons that they are popular and have an advantage in certain feedlot areas.

How Are Holsteins Different? Let Me Count the Ways

- Playful and gentle temperament for steer, but unpredictable and dangerous if accidentally left as bulls.
- Easily bored and therefore they "sort" their feed which can lead to metabolic problems.
- More bloats and "metabolics."
- More bullers.

- More running and playing which means more dust in the pens.
- Feedlot cowboys may not like them because it's hard to chase an animal that insists on following you.
- Heat tolerant, but cold intolerant.
- At risk for liver abscesses and acidosis, but seldom founder.
- Holsteins drink more water than beef cattle so their pens may get muddier.
- Suicidal tendencies

In the past several years, sexed semen technology in conjunction with AI programs has become more popular, and there has been an over-production of dairy heifers. Perhaps we could use sexed semen technology in reverse by selecting desirable semen from beef sires to produce Holstein cross-feedlot steers. They've not only got good performance characteristics but also excellent carcass quality. One might assume that future dairy expansion may move to areas with the greatest feed availability that would potentially match where the cattle are fed.

If the expansion of Holstein feeding occurs in the Midwest, it most likely will be in confinement. The confinement feeding concept became popular for a short period in the early '70s. Whenever anyone would contact me, I would point out that not only were they costly, but they had some shortcomings in cold climates. Nonetheless, people tired of mud were willing to try anything and the designs were modified in various ways but not necessarily improved. There were teardrop designs, flume floor designs, increased solid concrete areas and steeper sloped pens. What each of these designs had in commons were more concrete and less slots and that is exactly what one didn't want. The greater the concrete surface, the colder the cattle were in the winter. Probably the worst design were flume floors which consisted of solid concrete slabs of approximately four feet that were placed on angles. In the low place in the middle, a flume for drainage was constructed. With this concept the cattle were always standing on an angle and almost always on solid concrete. This design was a performance killer.

There are potential advantages to close confinement when you can keep cattle out of the mud and minimize the cold stress. There are new designs such as hoop barns, monoslope buildings that are apparently working well. Consultants McClellan and Peters have done a lot of work on new confinement designs. Also, with higher fertilizer prices, there is a real advantage to collecting and conserving the manure and to spread on adjacent farm ground. In some instances, we have estimated that the manure saved may contain as much as thirty to fifty cents per head per day fertilizer value if properly handled. In the long term, feed follows water and cattle follow the feed. If feed production favors the Midwest, cattle will move to that area. Furthermore, Midwest weather may favor more confinement developments.

-CHAPTER 13-

Nevada—It's Not Just Bright Lights

As a kid, for some reason I always had a fascination with Reno and Las Vegas, Nevada. Sometime in the mid- to late-'50s. I jumped on a bus and traveled to Reno just to see the sights. I didn't tell anyone my plans, and ended up spending about a week in the Reno, Virginia City, Las Vegas area. It was a real eye opener.

I arrived at the Reno bus station about 6 AM. A lady in a bright red dress and her girlfriend who had been on the bus introduced themselves asking, "Are you a cowboy?"

I answered, "Sometimes."

The lady in red said, "Great, my friend is from Germany, and dying to meet a cowboy. Why don't you spend some time with us."

I said, "Okay," but thought they must have something devious in mind. I was green and suspicious.

The lady in red had a friend who drove up in a new purple Cadillac convertible and said "Get in, I'll take you to my casino."

We went to Virginia City where he owned a small casino called the Delta Saloon. We had a drink, and he gave me his car keys and telling me, "Take the ladies with you then pick me up around five." We spent three or four days touring the various casinos and ghost towns. I had one of the best times of my life. The casino owner and two ladies became my good friends. When we parted, we promised to stay in touch, but suspicious me had given them a false name and address and was ashamed to admit it. I went from Reno by bus to Las Vegas. My first stop was the Golden Nugget

Casino. They had a live band playing in the cocktail lounge. It was the Billy Gray Band who sounded like Hank Thompson.

A lady sitting next to me said, "I love this band, but the band leader keeps hitting on me. If you don't mind I'm going to tell him you're my husband when he comes by." Sure enough he stops by the table and she introduced me as her husband. My new friend showed me several casinos, and by then I was worn out.

Then she inquired, "What are your plans." I explained that I've planned to take a 6:30 AM bus the next morning.

She said, "You'd better come with me because you'll get in trouble by yourself."

I thought, "Wow, what's next?" What was next is she made me sleep on her couch and took me to the bus terminal at 5:30 AM the next morning. That ended my first week in Nevada.

Ironically, about fifteen years later one of my first clients was in Nevada. I received a call from a Hollywood business management firm who asked that I meet them in the Reno-Carson City area to look at a potential feedlot site for Tennessee Ernie Ford. We went to a beautiful ranch that lay on the Carson River. Ford had bought it from the bankrupt estate of Charlie Steen. Steen had made the first big uranium strike in the United States. He was an old, hard-rock miner, and although he became a multimillionaire for a few years, he soon spent all his money and went bankrupt. The reason they thought it would be a good feedlot site is that it had four Harvest Queen Silos plus some feed bunks. The fact that there was no grain produced in the area, no packing plants, and so forth wasn't important. This was the era of eighty to ninety percent max tax brackets for the high earners who were consequently interested in the feedlot business as a tax shelter. Ford was also interested in marketing private label beef that he planned to name Tennessee Ernie's Pea-Picking Beef. He played the Reno–Tahoe area casinos frequently, and had ranches in Oregon and Northern California. He liked cattle in general.

A couple of months later, the Ford group called me at the Veribest Feedyard in San Angelo and asked me to be at a San Francisco bank meeting first thing the next morning. They were putting together a feedlot loan. I flew all night, and when I hit the San Francisco airport, they'd lost my luggage. I was in blue jeans, a western shirt and boots. The only clothing I could find at the airport gift shop was a Hawaiian shirt. I went to the top floor of the Bank of America offices in San Francisco dressed in blue jeans, boots and a Hawaiian shirt. They all looked at me and said, "This is our feedlot expert?"

The bankers and the Tennessee Ernie Ford group argued all morning about the size of the loan, repayment, and other details. About 11:00 AM we adjourned to a club for a three-martini lunch. Subsequently, everyone was mellow, and we put the deal together in a matter of minutes then went our separate ways. I continued to tell them this was a bad spot for a feedyard, and that they had under estimated the cost to construct a feed mill. We did bring in the Williamson Built Mills people from California to give us a bid and after this they lost interest.

I had noticed the large silos had an added attraction of flashing lights on the top of each silo. An old Indian who had previously worked for Steen told me that Steen made a deal with the Madam of a bordello next door. He put the spotlights on top of the silos as a landmark in exchange for free services at the bordello.

A final note to this story was the last time I went to the ranch I sat next to a fellow on a flight from Phoenix to Reno. He was the president of a silver mining company who had a mining claim near the Ford ranch. At the time Howard Hughes was buying a lot of land in Nevada and everyone was convinced he knew about great new silver deposits.

The mine president told me their location saying, "You know

it's next to a bordello, and the lady who runs it is shooting at our men on the heavy equipment."

When he reported it, the local sheriff said, "I know that old bag. Just shoot back."

The president said, "I don't want to do that, so I'm up here for negotiations with the lady. I believe a group of miners and a bordello should have some common interests."

Shortly thereafter I had a call from the Howard Hughes organization and his ranch manager. They had a ranch northeast of Las Vegas with some unusual cattle problems. I asked him to describe the problems and the ranch. His reply was, "You won't believe it until you see it."

We drove northeast of Las Vegas towards a small town named Moapa near an Indian Reservation. We went over a desert mountain, entered a valley that was all green grass and palm trees waving in the breeze. It was a spring-fed oasis in the middle of the desert—an incredible sight. I diagnosed the problem with the cattle as a probable trace mineral deficiency, and gave them a mineral formula. That's the last heard from them.

However, I discovered that the second week of December was the slow time in Vegas, and you could get almost anything you wanted at bargain prices. This was prior to the National Finals Rodeo that solved that slow period problem. I had a combination early Christmas party and symposium for clients for a couple of years in Vegas. Later Jack Linkletter and I put together an international cattlemen's expo there. It was moderately successful, and I became acquainted with several hotel and casino managers who could always get me the best tickets for shows and dinner. My favorite hotel at that time was the old Sahara which is now defunct. The Sahara manager's father was the Number Two man with the Howard Hughes organization in Las Vegas. We discussed their cattle deals plus other points of interest. What was fascinating is that even though he was the Number Two man, he had never met

Hughes, and only talked to him a few times on the telephone. I've been told that Hughes' Number One man, Bob Maheu, only met Hughes once. Hughes was making a big splash in Vegas because he was buying up so many casinos. Everyone thought he was eccentric but brilliant. That eventually proved to be only half true.

There was another thing that made Vegas interesting to me. When I was teaching the Feedlot Management course at Texas A&M, Benny Binion, the legendary Texas gambler and owner of Horseshoe Casino called to ask if I could get his son-in-law into my class. Because it was a graduate level class, I told him I didn't think he would qualify academically, but he could audit the class and I'd make sure he got everything the regular class members received. Benny never forgot a favor. Consequently, whenever I came to Las Vegas I would stop to say "hello." Although I didn't meet him often, I was always impressed because he would recognize me instantly. Benny was a colorful fellow whom Amarillo Slim once described as "either the gentlest bad man or the baddest gentleman you will ever meet."

At that time, Andy Michaels was married to Benny's daughter, Brenda. He ended up managing a feedlot for one of my clients in the Panhandle. Brenda, who lives in Amarillo, had a sly but great sense of humor. I called her one day from the Amarillo Airport to ask if she had time for a drink at the airport bar. She said she thought she should steer clear of the airport. Recently she had accidentally forgotten a pistol in her purse, and it was taken away from her by airport security. She said, "I don't know what the big deal was because it was just a little pistol—but it was one of her favorites."

The atmosphere in Vegas has changed in recent years as it has in Reno. I don't travel there much anymore, but fond memories of bygone days linger.

Caroline and I often went to the National Finals Rodeo. When she was young she competed in barrel racing and was a rodeo queen. She tells a funny story about Elvis playing at one of their rodeos in the mid-'50s. After the rodeo, Elvis asked her for a date, however, her mother and father were along, and she not only had to stay in their room, but she had to sleep between them. Going to Vegas with Caroline at rodeo time was an adventure because someone was always asking for her autograph. They would say, "We don't know who you are, but we know you're famous and please sign this." Our standard joke each morning was who are you going to be today. Caroline's granddaughter, Casi, is active in barrel racing. Caroline always hoped she would make the National Finals but that hasn't happened yet.

A footnote to Benny Binion: he had the same type of sly humor as his daughter, Brenda. One time he, John Scott and I were having lunch in Billings, Montana.

John said "Benny, I hired a fellow who said he once worked for you. I want you to know he came to me, and I didn't try to steal your help."

Benny asked who it was, and when John told him, Benny's reply was, "That's okay, he's just a 'monkey wrench thief.'"

I asked what that meant, and Benny said, "He doesn't have the balls to steal anything big, but he will steal your tools." Benny also said the Depression wasn't too bad because chili was 10 cents per bowl and he liked it. After Benny died, I called Brenda and expressed condolences. I mentioned I'd read interviews with him in the Houston newspaper where he got very explicit about his personal life. Brenda said that the family knew he was sort of losing it. The last year he spent much of his time in his chartered bus travelling with Casey Tibbs who was dying of cancer. The newspaper interview really made her mother mad because Benny claimed her chain smoking killed their favorite pet poodle. The fact that he had also discussed a couple of gunfights and killings wasn't near as important to her as the poodle incident.

CHAPTER 14

Arizona and California—Birthplace of Modern Feedlots and Consultants

Hal Williams Jr.—A Personal California Favorite

It may surprise some newcomers now that the modern feedlot industry and consulting business began in Arizona and California and flourished until the wreck of 1974. Take a look at the seven-state cattle feed report in 1974 as compared to 2013. Obviously California and Arizona numbers have declined substantially, and the rate of decline may increase with the recent announcement of the Brawly beef packing plant closing. It should be noted that California is also the home of a lot of beef and dairy cows, and for a variety of reasons (such as drought, alternative land usage, state bureaucracy) these numbers have decreased.

Early vs Current 7 States COF (million head)

	Jan. 1974	Jan. 1975	Jan.2014
Arizona	609	319	274
California	1209	688	510
Colorado	938	755	960
Iowa	1715	1200	620
Kansas	1160	920	2010
Nebraska	1525	1160	2400
Texas	2205	1327	2440
	9361	6369	9214

The following are a number of old pictures taken by a Champ Gross who began *Calf News*. Barb Jackson provided these and they feature a number of prominent old-time Arizona and California feedlot people. Barb's father, Carl Stevenson, who is a pioneer feedlot man in Arizona is 96 and alive and well. The feedlot is now managed by his son, Dave.

Interested feeders at the Artesia delivery point were: Joe Finley, of Western Cattle Feeders, Blythe; Bert Byrd, Janss Cattle Industries, Thermal; Jack Torpey, Torpey Cattle Co., Los Angeles; Robert McLeod, Rummond Bros., Coachella; and Robert McMicken, of the J. G. Boswell Co., Los Angeles

Prospecting for his Sweetheart

K. S. Eng

There's a desert town named Glanmis
Where the railroad use to stop
They had a Post Office and a bar
Cramed into it's only shop

It's on the road from Blythe to Brawley
Sometimes I would stop for
a few beers
That's where I met a lonely miner
That I've remembered for 40 years

He was sitting alone at the bar
Sipping a semi-cold brew
I introduced myself and he said
"I'm glad to have someone to talk to"

He wore faded jeans and a levi shirt
Said he was prospecting for gold
He needed a stake for his sweetheart
He was searching for the Mother Lode

"I'll show you a letter from my girl"
And a faded letter he began to unfold
It said she'd wait for him back East
Hoped he'd return with pockets of gold

He carefully refolded the letter
On the envelope I noticed the date
It was postmarked May 1939
It was now May of 68

His pale blue eyes grew misty
Said he hadn't heard from her for a spell
But hoped she'd wait while he
kept searching
For gold in this desert hell

I often think of that lonely miner
And how hard he hoped and tried
But for good luck and good friends
We might have side by side

Arizona Feeders

Ed Tovrea

Sam McElhaney

Chuck Chritchlow and Carl Stevenson

In the mid-'70s, a lot of Arizona and California feedlot managers moved to the Texas Panhandle. They weren't always welcomed with opened arms. I know one feedlot manager who had a prominent sign in his office that said "We're in Texas, and I don't give a damn how they do it in California and Arizona." In general, however, the Texas feedyards were quick to embrace the West Coast technology provided by feedlot consultants.

One of my first independent consulting clients was Hal Williams, Jr. of Williams & Sons feedyard near Lake Matthews, California.

It sat in the middle of orange groves, and was truly gorgeous. Hal was a great feedlot operator and a beautiful man. He mainly fed light, crossbred calves from the southeast or calf-fed Holsteins which were almost entirely marketed to Vernon Street Packers, and a packer near Escondido, California. His ideal slaughter animal weighed 900-950 pounds and the conversions were great.

One of my habits at a feedyard is to walk a number of the pens, and I often do it over the noon hour when there is limited feedlot activity. I was walking a Williams' calf-fed Holstein pen when I heard a low bellow behind me and turned to see a Holstein bull weighing about 1000 pounds charging me. I grabbed him by his horns which were approximately 4-6 inches and turned him so he didn't hit me in the gut. He still slammed my side as I wrestled him down, jamming a horn into the ground. Then I made my worst mistake—I kicked him in the head. As I walked away, he followed me, hit me again, and continued to knock me down several times, but each time I was able to grab his horns and turn him so he didn't hit a vital spot. I was just about worn out when the secretary came back from lunch and saw the commotion in the pen and jumped over the railing to help me. The Holstein bull took after her, but she was able to make it out of the pen, and I was able to crawl out on my own. We went back to the office where I was sore as hell. I realized I had two broken ribs and lots of bruises.

When Hal returned from lunch, I asked him, "Have you got a gun in the office?"

He said, "Yes, a .30-30 Winchester."

I told him, "I want it, and I want to buy that Holstein bull so I can kill the SOB. I want the pleasure of shooting him a couple of times in the gut."

Hal said "You're welcome to him, but he's scheduled to go to the packing plant tomorrow morning. If you can wait, you'll save about $500 and he'll die anyway."

It turns out that earlier that morning the bull had attacked a cowboy on horseback and knocked both him and his horse down which illustrates the potential danger of a one-nutted Holstein bull in the feedyard.

Hal and his wife, Helen, were friends with me and my family. Unfortunately she died of liver cancer at a relatively young age. Hal was a handsome, vigorous young '70s fellow who had lots of lady friends in church that were determined to take care of him. One night I was in the airport in Des Moines, Iowa when I was paged. The message was to call Hal Williams immediately. Fearing the worst, I returned the call.

Hal informed me, "Kenneth, I've got a big problem and I need your help. I've got one woman in the kitchen cooking dinner, and another measuring my drapes in the living room. What do you think I should do?"

I retorted, "Hal, why are you asking me?"

He said, "You've had more experience with women than I have, and you're my consultant."

I informed him, "I'm not that kind of consultant, and besides, at least you don't have someone bringing you a casserole.

He replied, "Oh yeah, I forgot to mention that, a lady with a casserole is coming later this evening."

Hal remarried and had several more productive years, but he also died of cancer at a relatively young age. I attribute part of this to the fact that we had a liquid diethylstilbestrol concentrate tank that was metered directly into the feed mixer. Hal would occasionally put his arm in the liquid diethylstilbestrol tank and stir it around. I told him not to do that. He responded, "Why? It doesn't hurt." I suspect it did. An interesting footnote here is that their original feedlot is now occupied by Los Angeles International Airport. Time marches on.

Hal Williams

-CHAPTER 15-

Vet Consultants—Friends Not Foes

A lot of lip service is given to the competition between veterinary and nutrition consultants but personally, I've never experienced a problem. My theory is that a successful vet consultant is busy enough doing vet work, and the same can be said for a nutrition consultant. They have no reason to overlap or compete. My advice is that if you have a vet consultant who claims to be competent in nutrition or vice versa—beware.

I've known most of the original cattle industry vet consultants such as Cerniga, Sheldon, Brimhall, Rinker, Ed Johnson, Dehyle, and the current crop including Nofsigger, Griffin, Del Miles, Dehyle, Jr. I'm leaving a lot of individuals out due to lack of space and poor memory. A lot of friendly heckling goes on between the groups. Del Miles once said, "The advent of power windows in pickups cut the nutritionist's workload in half." I told Del, "The reason veterinarians talk incessantly about bloat problems caused by nutritionists is that veterinarians can't spell entrotoxemia." Del and I have the same birthday, and we think we may be twins—but not identical.

Of course you can't discuss veterinarians without mentioning Baxter Black who truly is in a class of his own. I was nutritionist for Simplot where he was employed at that time. At that time, Baxter and his wife weren't getting along. but he subscribed to the old Willie Nelson School of romance which says "if it feels so bad, it must be love." He has since remarried a great lady. He and Cindy prove that "love is lovelier the second time around." He's a close, personal friend and he preached at Caroline's funeral. Baxter has become a religious soul, and I now call him "The Reverend Dr. Black." He's a great entertainer, and even better human being.

Another of my favorite vets was Hal Rinker who was one of the original Southern Plains vet consultants and a pretty colorful fellow. Hal developed a severe drinking problem and personally told me that when he got up in the morning, he had to drink a pint of vodka before he could get started. He would continue to drink all day by filling the windshield washer of his car with vodka. He disconnected the hose from the wiper, put the hose in his mouth then pushed the washer button whenever he needed a drink. He said that by the end of the day, no one could smell the alcohol, so they just assumed he was stupid. Hal "took the cure" and was one of the lucky ones who quit drinking and maintained a good sense of humor and personality. A lot of us have encountered those who quit drinking but became so cranky you often wished they would sneak a sip. Another great thing about Hal is he counseled many individuals as well as penitentiary inmates on substance abuse problems. Hal died this past year and I'm betting there's a special place for him in the big sick pen in the sky.

-CHAPTER 16-

Computers and Computer Systems

I was never very good with math or hardware, but when I saw the first electronic portable calculator in 1968, I had to have it. It was about the size of a portable typewriter, was made by Olympia and cost $795. The major attraction for me was that it couldn't be jammed which was a constant problem I had with older machines. I bought it from the local office supply store in Casa Grande, Arizona and told them I was worried about upcoming changes that might make it out of date. They suggested I put it on a three-year payout plan, and if something better came along I could apply the payments to the cost of the new unit. About a year later, someonce next to me on an airplane pulled out a Sanyo calculator which was about four inches by six inches and two inches thick. It could do everything my Olympia did. It only cost $250. When I traded my Olympia for the Sanyo at the Arizona office supply store, they owed me $200 for the trade. My old Sanyo still works. Of course, now you can buy a pocket-sized calculator that will do as much or more and probably cost $3 to $5.

In early 1969 Harry Konen, who was a poultry consultant from Houston, Texas, had bought a new least cost formula computer unit called Scidata. It was a remote unit operating online to a central computer Honeywell 2200 in Atlanta. It was a fascinating little unit, and at that time had many new bells and whistles. I went to the home office in Atlanta to learn more about it. I met a small group of fascinating people, and got involved with their least cost programs. I was asked to build a feedlot computer performance projection program. The president and brains of the group was a large fellow who was brilliant. Also helping on their least cost program was a well-known poultry consultant from the Delmarva area named Perry Twining. Their main salesman was a fellow named Bill Benge from Texas. The least cost ration format Twining and I put together is still similar to many current programs.

The feedlot performance projection program is very similar to many currently used. I designed it based on the California net energy system developed by Lofgreen and Garrett. For my money the California net energy system was the most useful tool developed for the feedlot industry in the '60s. Garrett and Lofgreen were great scientists. Glen Lofgreen became a personal friend when he moved to the Clayton Research Station. I still consider him to be one of the best university research nutritionists I've ever known. I stopped to visit Lofgreen the day he officially retired from New Mexico State University. He was sad and mad as hell because New Mexico State was forcing him into retirement because of an age limit. It was a loss to the whole industry because he was still doing great work.

Back to my feedlot performance projection model, it is still the basis for many feedlot printouts today. I designed it from scratch, but I thought there might be a similar one floating around somewhere. When I checked in with people whom I thought might have had one prior to me, they told me "no," they had copied mine. Unfortunately, I neglected to copyright the printout format. Considering the number of these printouts generated every day that cost me some big bucks.

In addition to hard work, we had a lot of fun in Atlanta. The first computer units sold to feedlots covered about half a wall each and cost between $125,000 to $150,000. They were successful, but Scidata had some internal problems in Atlanta. Among other things, the President of the company had narcolepsy. I'm told he fell asleep at an early stockholders meeting. Needless to say, the stock price crashed. As part of my payment, they gave me stock. I sold some of it for $1,000 per share prior to the narcolepsy event. I wish I had sold it all.

I stayed at the Peachtree Plaza, a fancy new downtown hotel in Atlanta where there was lots of excitement and entertainment. One morning while I waited to be picked up, a fellow in a uniform drove up in an old station wagon, got out and yelled, "Airport transportation." A bunch of businessmen waiting to go to the airport started to pile in when he said wait until I get your luggage and then we'll match the people to the luggage. He proceeded to fill the back of the station wagon with luggage and drove off leaving everyone dazed and yelling for the police.

Perry, Bill and I were always coming up with weird ideas. One of us mentioned that there were a lot of Naugahyde® couches and chairs, but no one was growing naugas. We decided this would be a good business so we formed the Society of International Nauga Scientists or SINS Inc. We put together a brochure and even ran a few small ads offering stock. Unfortunately, we started getting investor calls so we decided we'd better terminate the project before we got in trouble. I guess the moral of the story is you can sell anything if you have a good story and a catchy name to go with it.

One of the last times I saw Glover was in the 1970s in Vegas where we met because he loved to gamble. He was a good card counter at blackjack especially back when they were using only single or double decks. He knocked on my door about four in the morning and said he had to leave for the airport right away. I asked him why he couldn't wait and have breakfast. He said, "You don't understand. They caught me counting cards and hotel security is escorting me to the airport, pronto."

Their Scidata computer system also worked well, and we still have one in Dale Furr's old office in Hereford, Texas. It's orange and takes up about half the wall and I think Bob Beville goes in and uses it once in a while. I bought it from Foster Farms in California in the early 1970s when they were going to replace it with a new system.

They gave me a good price— about $15,000—for the equipment. When their new system didn't work they had to lease it back from me for almost a year. The lease payments were greater than what I paid them for the computer so it was obviously a good deal for me. I gave half-interest in the computer to Dale Furr when he started his consulting business, and we shared it for many years thereafter between gin and domino games.

I overdosed on computer work early. Computers now hold little interest for me. Obviously, this puts me at a current disadvantage. My major problem with computer technology is it has dumbed-down oral communications. I'm still big on personal or phone conversations when doing business. As Larry McMurtry said in his book, *Duane's Depressed*, "Business is not complicated; just return your phone calls."

-CHAPTER 17-

Tax Shelter Feeding

No discussion of the 1970s would be complete without discussing Tax Shelter Feeding programs. Maximum tax rates at that time were in the 70-90% range. Shelters varied in sophistication and complexity, but the basis concept was to buy a light calf, and prepay the entire feed bill of approximately $150-$200. The prepaid feed was tax deductible. Furthermore, financing for approximately 75% of the calf and all of the feed was usually available. For $25 per head cash, the tax deduction result might be as much as $200. Some of the financing was nonrecourse. A number of cattle feed programs in the early '70s crashed and burned because of the drought and the market crash in '74. To be fair, the same could be said for feeding programs that were not tax shelter oriented. When times got tougher, participants were more desperate, and often their management had no real concept of the cattle business. The poem I wrote entitled "Phantom Cattle" is a pretty accurate portrayal of what happened with some of the programs.

In late December prior to the year's end, the tax shelter business usually peaked. Between Christmas and New Year's my phone rang constantly with clients asking for projected feed usage, break-evens and other information regarding various types of cattle. I thought it was exciting, but it didn't make me popular with my family during the holidays.

My clients, Jack and Art Linkletter, had a tax shelter investment fund called Agrilink. I worked closely with Jack. He was a great guy. Jack took good care of himself but he died at age 70 while his Dad died at approximately age 95 and looked great to the end. Goes to show you that you never know.

One of my jobs for Jack was to approve feedyards for their cattle placement. This aggravated their comptroller who was from East and had no real concept of the cattle business. He just didn't like putting cattle on feed without knowing the exact cost, thus he was always a sucker for any guaranteed gain deal. I told him that anyone who guaranteed the gains beforehand was suspect because good feeders didn't need guaranteed gains to get cattle. A large feedlot in Southern California was reopened around 1973 by a man with Hollywood connections, but he had no cattle experience. He contacted Agrilink and said he would guarantee gain costs for the placement of several thousand cattle. I convinced the Linkletters it was a bad idea. Unbeknownst to me, one of my other California clients whose feedlot was full took the guaranteed gain proposition. They needed a place for about ten thousand extra cattle. The first I knew of this was when my client contacted me saying we needed to go check these cattle. We got there and the feed bunks were filled with a mixture of straw and manure. We asked the feedlot manager for an explanation and he told me there was research that proved this was a decent ration.

I knew Jim Elam was doing their consulting work so I said, "This is new to me, but I'll call Jim and confirm it."

The manager replied "No don't do that, the real reason is the owner resold all of our good grain contracts, and now at higher grain prices he can't afford to buy grain."

We moved the cattle and it ended up in a big lawsuit which my clients won. A footnote here is that some years later I was told the defunct feedyard owner was found stabbed to death in an alley in the Central California Coast area. I often wondered if this was a random act of violence or some type of payback.

Another California tax shelter program got in serious trouble in 1974. They put feedlot cattle on manure and straw diets, and had starving cattle scattered throughout southern California. PGC of Amarillo owned a large grain elevator in Colten California, and the manager of the troubled feeding fund owed them for several thousand tons of grain. He also owed several California suppliers money. This is where it gets interesting. I was doing the nutrition work for the PGC feedyards in the Texas Panhandle where Moody Taylor was the manager. Taylor was, to say the least, a very clever person. With approval of the feeding fund, he moved about twenty-five thousand cattle from California to the PGC Texas feedyards before any California creditors knew about it. In a way, it was a humanitarian gesture because the cattle were starving, but obviously the California creditors were outraged. PGC agreed to finish the cattle in Texas, sell them and deduct the cost of the Texas feed plus the feed they were owed from California. The California creditors and investors would get the remainder. In the end it was one of those programs that was "snake bit" from first to last bell. We had a bad winter that the thin Holsteins didn't tolerate well in Texas. When the final cattle were sold, they were bought by American Beef. About two weeks later American Beef took bankruptcy and PGC assumed they had dodged a bullet and got in under the wire. Unfortunately, the bank did not clear the check promptly and it became part of the bankruptcy proceedings. I understand that eventually the bank was forced to make the check good. Regardless, a lot of lawsuits resulted from the original feeding fund investors and creditors that continued for the next five to ten years.

I don't want to leave the impression that tax-deferred feeding never worked because sometimes it served investors well. Around 1980 I put together a cattle feeding program for investors in the seventy to eighty percent tax bracket. The cattle made some money and we deferred their money over the next several years into a forty-five percent tax bracket. It was obviously a net winner.

One last comment on Jack Linkletter. He was a bright fellow, fun to be with and a pioneer innovator in many respects. Jack told me that his grandfather said, "You can always tell a pioneer because he's the one with a lot of arrows in his ass." After the mid-'70s wreck, Jack felt a lot of arrows in his backside.

Phantom Cattle

K. S. Eng

It was late in 1973
Tax shelters were still the rage
But the cattle wreck had started
We were about to turn the page

Corporate lawyers &, accountants
Were it a loss for words
While investors & their bankers
Kept searching for phantom herds

When short on numbers excuses varied
Like too much rain & the grass
is too high
The cattle strayed when fences
washed out
Some of them drown &
some always die

Or, high country cattle are always
wild as hell
With water they're on the
highest mountain
They'll come back when the
tanks are dry
Next year we'll be able to count them

Near the end excuses became
more exotic
Most of the tellers weren't cattlemen
My favorite was about the
herds, so wild
That there were no cattle left in them

While bankers tried to count
phantom cattle
Some corporations had a
trucking branch
To make up numbers the trucks came
in handy
Hauling the same cattle from
ranch to ranch

When it was over there was some
good news
Because when the final story was told
The trucking company made
lots of money
And the cattle knew how to load

Those involved have all scattered
You can find them in most states
Some are involved in telemarketing
Some learned to make license plates.

-CHAPTER 18-

Expert Witness at Lawsuits (A Hard Way to Make an Easy Living)

I have probably served as an expert witness in twenty to thirty lawsuits. Some were for clients, but the majority were for other groups who believed I had information that would help them in their lawsuits. It started well since I picked my clients and lawsuits carefully, and I believe early on we had ten lawsuits where I gave depositions. They were all settled prior to trial and in favor of my clients. Needless to say, I became a little cocky and over-confident thinking this was an easy way to pick up some extra bucks with a minimum of effort. Unfortunately, if it seems to be too good to be true, it probably is.

In the mid-'70s I was an expert witness on three lawsuits that went to trial simultaneously. One was in Montana, one in the Imperial Valley and the third in the Texas Panhandle. You learn as an expert witness that the lawyers are not concerned about your schedule, and the judges really don't give a damn. For a month I spent the majority of my time flying from one lawsuit to another. Even though my clients won each lawsuit, it was a real pain in the butt. You reach the point where you never want see another lawyer or judge and they probably felt the same way about me.

The Imperial Valley trial had to do with an honest-to-God Vitamin A deficiency on a very large scale, and it was the first and last I've ever witnessed firsthand. The defendants denied the responsibility for the Vitamin A deficiency, but it was determined that the Vitamin A supplements they supplied were deficient probably because of the effect of extreme heat on Vitamin A activity. The defendant's lawyer was loud but small. One night at a hotel where we were both staying, he got into a serious altercation at the bar. He was seated near me and a couple of Hispanic men. They decided he was a smart ass gringo, and they were going to mop up the floor with him. I intervened on his behalf and convinced the two Mexicans it wasn't worth their effort and they left. I became the opposing lawyer's new best friend. I don't know if this had an impact, but the lawsuit was settled in our favor soon thereafter.

The trial in the Texas Panhandle was amusing because my client's lawyer informed me prior to my testimony that he didn't think much of expert witnesses. He could win this trial without me although I would probably be called. The trial dealt with misrepresentation of the quality of some roughage purchased along with a feedlot purchase. Both the hay and silage were misrepresented. I can't recall if I was on the stand, but our lawyer got up in front of the jury with a burlap bag full of moldy hay. He threw it across the table and said, "Would you feed this crap to your cattle." After the dust and mold settled, the answer was obvious, and he won the case beating an opposing team of several high dollar lawyers with a high profile client.

Another humorous incident occurred was when I was an expert witness for Gary Weisbart who was being sued by a customer because of unrealized fat cattle profits. Gary had actually made quite of bit of money for the plaintiff, but he thought he should have made more. It turns out he was basically a professional litigator and we went to court. Those of you who know Gary Weisbart know he never backs down from a fight and the case went to trial in Denver, Colorado. The day I was to appear as expert witness, I told Caroline if she wanted to go, be sure to keep her mouth shut because the job of the opposing attorney was to make me look stupid, and he would try his best. Also, I told her to dress down instead of wearing her furs or diamonds because she wore accessories worth more than a lot of the jury members would earn in a year. Caroline pouted for a while but agreed and off to court we went. Just after she sat down in the balcony, Gary Weisbart, the defendant, yelled, "Caroline come sit with me." Gary had on one of the gaudiest fur coats you have ever seen plus alligator boots. Caroline came up to me and said, "Why can Weisbart dress like that when he's the defendant, and I've got to dress like a camp cook just because I know you." It all ended well because my cross examination turned out to be a piece of cake, and the lawsuit was settled in Weisbart's favor. Caroline, however, was still aggravated about her dress code.

In the early '70s, two friends of mine got into a lawsuit because they were both hard headed and neither of them wanted to back down. One owned a feedlot near Chandler, Arizona and the other was a cattle trader. The cattle trader had put about 1500 heifers in the feedyard to be grown on a limit-fed, low-gain basis. He thought it was costing too much to feed them, and wanted to remove the cattle, but the feedlot owner wanted his bill paid first. The cattle owner removed the heifers on a weekend when, coincidentally, no one was at the feedlot. He told me, "I just happened to be driving by with some eighteen wheelers and no one was there, so I took the cattle."

My friends both hired high-powered lawyers, sued each other and each asked me to be their expert witness. The lawyer for one group was the criminal attorney who had gotten the original Miranda decision requiring a defendant be read his rights when he's arrested. One of his staff called me saying, "You're scheduled as our expert witness when the trial begins in two weeks." I let her know that I had already informed both parties I would not be a witness for either side. She answered, "We may have to subpoena you." I thought how in the hell can you subpoena your own expert witness, but her comment stuck in the back of my mind.

Pat Ishmael was my office assistant at that time, and she had some legal experience. When she came into the office one morning she told me, "There's a seedy-looking fellow in an old car circling the block. I will bet he's going to try to serve you the subpoena." Sure enough, he rang the doorbell a few minutes later, and asked to see me. He had a summons in his hand. Pat told him I was out of town, and wouldn't be available. However, he continued to hang around for a couple of days. I just stayed in the office or bedroom or by the swimming pool. When the trial began, I was the only witness called other than the two involved in the lawsuit. Of course I was absent. My cattle trader friend told me the judge listened to him and the feedlot owner plus their lawyers then concluded it was a frivolous lawsuit. He told them to come back in a week with a proposed settlement or he would fine and possibly jail both litigants and their lawyers. Meanwhile, I was suffering from cabin fever, so I hid on the floor of the back seat of the car and had Pat drive me to the Phoenix airport to fly to New Mexico.

Perhaps the most aggravating lawsuit I was involved in was the result of an acquaintance claiming in their feedlot brochure that I was their nutrition consultant. A group was suing his feedyard without a good basis, and they called me because I was listed as the nutrition consultant. I replied I was not their consultant, and I then called the feedlot owner and asked him what the hell the brochure was about.

He said, "Kenneth, remember just before we opened the yard, we discussed my feedyard at a cattle feeders' meeting?"

I came back, "You know damn well that doesn't make me your consultant, but your brochure implicates me in your lawsuit."

He reply was classic. He said, "Well, if I didn't need you before I really need you now as my expert witness. Please help me and charge what you feel is appropriate."

I did, but it was a short-term aggravation.

The bottom line is I still receive calls asking that I help as an expert witness, but I guess I'm either smarter or lazy or both. I decline and refer them to friends. Perhaps the greatest downside in the expert witness business is that you're forced to associate with some people (usually the opposing lawyers) that you wouldn't in real life. One of the greatest advantages in my career is that I didn't associate with people I didn't like or with whom I wasn't compatible. In depositions, lawsuit testimony, and that sort of business, you are occasionally forced to tolerate an opposing lawyer who thinks it's his God given duty to act obnoxious. Perhaps it's part of their strategy to intimidate you or perhaps it's just part of their real personality. Maybe there's a special class in law school that teaches trial lawyers bad manners and how to be obnoxious. Law students should ask their tuition refunded because the strategy is often counterproductive. This is not to paint all lawyers with one brush, and as I have grown older and mellowed, we've used several good lawyers in personal business transactions. Most have done a good job for us, and I've even enjoyed their company.

-CHAPTER 19-

A Traveling Man
(The Curse of a Curious Mind)

I've always been a curious person with a lot of stamina. Add that to my desire to see new country, and you have a natural-born Traveling Man. I started traveling on business in the early sixties and traveled constantly for fifty years. I suspect my air travel would total four to five million miles. I've probably stayed in hotels and motels over 8,000 nights. With this much travel, it's not surprising there are a lot of stories to tell. The following are a few of the more exciting or humorous ones.

As a kid I was fascinated by history, geography and travel. However, in reality we seldom ventured beyond a fifty-mile radius of our farm. My travels took place in my dreams, the one-room country school or in the kitchen at night with Mom and Dad where we studied maps, and history books while dreaming of places we'd like to see. I might have majored in history or geography in college, but couldn't figure out what to do with those degrees. I was able to send Mom and Dad on several trips including a cruise to Alaska—a place Dad had always wanted to visit. Dad also was able to see my California ranch before he died of cancer at the relatively young age of seventy-two. He had a good life but cancer was eating him up so it was time for the final bell to ring. Mother traveled with me occasionally, and while she liked my friends, she would remark that they were interesting but some seemed a little wild. I told her that's why they're interesting. She was never sure I had a real job because I didn't have a normal office, a regular schedule or a boss. Once she was flying back to Nebraska seated next to a young fellow who worked for a large feed company. They began talking and he said, "Are you Dr. Eng's mother?" When she answered "Yes," he told her that he would give anything in the world to have a job like mine. He later told me that she responded, "I appreciate that, but could you tell me what he does?"

Until I was eighteen, the only out of state trips we took were once to the Black Hills and Badlands of South Dakota and once to Estes Park, Colorado. We also made a couple of trips back to Illinois to visit Mom's relatives. My cousin, Sally, and her husband, Willis, still live south of Decatur, Illinois when they're not wintering in Florida. Cousin Sally is an excellent writer and has written a great book entitled *Lessons Learned from Common Sense* that I highly recommend. Our Illinois relatives and friends would visit each fall during pheasant season. Because Illinois had been spared the ravages of the dustbowl and they could afford a yearly trip I thought they were rich.

Not only did I seldom venture beyond Nebraska, I didn't get to Omaha or Lincoln until I was thirteen. We didn't travel far, but we wore out the roads within a 50-mile radius of home. When I started college at Wayne State in Nebraska, that expanded my horizons a bit more. Yankton, South Dakota became a favorite out-of-town party spot where the legal beer drinking age was eighteen.

As I mentioned earlier, when I was eighteen I worked in Florida after the mid-'50s drought set in, then went to the Catskills in New York the following summers. After that, I never stopped traveling. My motto during my early consulting career was "Never pass up a chance to see something new and see it firsthand." I often did it 24/7 which obviously was the game plan of a fool and put a strain on family life. Charlie Andrews told me the cattle and consulting business is great for men and mules, but hell on horses and women. I guess I'm living proof of this as I wore out several good women. The best I can now do is apologize and say, "I didn't mean to be unkind." Part of the magic of my relationship with Caroline was that she could keep up with me every step of the way. She was just as curious, thus we were perfect partners. Many people think they would like to travel but after a few years burn out. Caroline and I never did.

My curiosity also included unusual cattle, and since I've probably seen 25 million cattle in my lifetime, there are several unusual groups or individuals amongst that population. One of the most interesting and wildest groups of cattle were some large white Brahma-influenced steers from Florida that we brought to the Rio Grande Feedlot at Eagle Pass. They came from a large Florida ranch that never allowed testing for brucellosis so they never sold any females. But they did castrate the bull calves and shipped them whenever they could find a buyer. Usually the cattle were shipped at 800–1000 pounds and the price was $10–$20 below the market because there were few repeat buyers. The first time we bought these cattle it was evident you couldn't run them through a chute and you never wanted to put a man or horse in the pen. Also you needed to add a couple of extra rails to the pens they were in. In spite of this experience, Rio Grande Feedlot bought the cattle a second time so they could go to Florida and watch them be gathered. They said this was a sight to behold because they gathered the cattle out of the swamps and palmettos with dogs alongside jeeps with cow catchers on the front. Any steers the dogs couldn't bring in were run down with jeeps, tied down and dragged. I'm told an Arizona trader bought these cattle once and made a deal with the Montfort Feedlot to partner and feed them. Neither the cattle trader or Montfort's had seen the cattle before. When they arrived, several steers were killed trying to send the cattle through the chutes.

Another group of wild cattle were some Arizona desert steers that had originally been Mexican fighting bull stock. This was also a group of cattle that wouldn't allow a man or horse in the pen. One Sunday they were shipping a load to the packer, and the truck driver came in with a single deck truck but he drove it under the double deck chute. He opened the chute gate and they ran up and jumped off on top of the truck. The manager came by saying it was the first time he had seen those cattle stand still as they looked around from the top of the truck. Then the roof began caving in and it all went south.

Occasionally you'll see individual cattle that gain weight so rapidly it's almost unbelievable. We had a large Charolais-cross steer that came from a wheat pasture 200 pounds heavier than the rest of the pasture cattle. When they arrived I said, "Let's weigh him separately, and weigh him again when they're shipped." During a 140-day feeding period, that steer gained 7.25 pounds per day. I had a personal steer calf from our New Mexico ranch that we shipped with a load of calves to Clayton, New Mexico. This calf couldn't have been more than 250 days of age but was very muscular and looked like a bull. When the feedlot called me with the weights they said, "You've got one bull in the group, what do you want to do with him?" I said, "If you'll check, you'll find he's a steer, and I want to know his weight." He weighed 950 pounds. I instructed them, "Sell him early because obviously he's going to be a fast gainer and I don't want a heavy weight dock." We sold him after 100 days on feed weighing in at 1550 pounds at approximately one year of age. I never had another calf close to that weight during the years I owned the ranch.

On the other end of the gain spectrum, we ran a lot of steer and heifer calves on wheat and summer pasture about 100 miles northwest of Fort Worth. We usually started with 350 weight sale barn calves. When we got ready to ship at the end of one summer, we had two well-marked steers that appeared perfectly healthy but only weighed 350 pounds. We did what any cattle trader would do and ran them back through the sale barn. They brought a good price as lightweight calves. The following year we had two more calves that looked identical to the previous two, and they also weighed around 350-375 pounds at the end of the season. The calves gained almost nothing during the grazing season even though they looked perfectly healthy and normal. It finally occurred to me some people were raising midget cattle for pets and I suspect these were a cross with these midgets.

A Southern California feedlot client had a Mexican steer with front feet like a horse rather than a split hoof. Every month, I'd look at that steer with the manager and we'd comment how unusual it was. But guess what, he went to the packer and neither the manager nor I had taken a picture of him.

When it comes to doctoring sick calves, it's not unusual for a cowboy to have a "favorite home remedy." Caroline and I had some overflow calves at a preconditioning yard managed by an ex-bull rider who often started his mornings with the breakfast of champions (Bud Light). He was a pretty good hand and good entertainment. One morning when we were looking at our cattle, he pointed to a steer calf that didn't look great, but appeared to be mending.

He said, "You know, I've had hell with that calf and did everything in the world to save him and nothing was working. Every morning he was looking like someone with a bad hangover. I decided my usual medicine program wasn't working so I'd try a hangover treatment and gave him 5cc's (IV) of Jack Daniels. He got up and was eating and drinking within an hour."

I told this fellow that he might be on to something, but to cut costs maybe we should try a cheaper generic whiskey like "Old Crow."

This is a horse story rather than a cattle story, but we had small palomino with great markings at a Florida dude ranch. All the kids wanted to ride him, but he had one bad habit. Whenever you would get near water, he would walk into the lake, lie down and roll over.

Fortunately, he never hurt any riders, but he gave them a bath. We usually saved that horse for spoiled kids. I called him the palomino preacher because he was always baptizing someone. I could go on with unusual stories, but some are best left untold and others are too much for everyone's imagination.

One of my first trips as a consultant was in 1963 or '64 for a Texas-Louisiana company called Red Barn Chemical. They were big in the fertilizer business and decided to get in to the liquid feed business. Problems with nitrate and urea toxicity were major topics at that time and Red Barn asked me to give a presentation at their annual sales meeting held in El Paso Texas. The meetings were held at the old Plaza Hotel across from the Del Norte—a cattle traders hang out. At the airport there was a limousine for the Plaza, and to my surprise, I was the only male. The rest of the passengers were women from New York. At that time the only grounds for divorce in New York was adultery, and the closest Mexican border town from New York City was Juarez across from El Paso. Many of the women were opting for Mexican divorces even though their legality was suspect. Not surprisingly, the El Paso cowtown and the Juarez culture were a shock to their system.

The Red Barn sales group was an exciting bunch, and coincidentally all of them had done business with the infamous Abilene business and con man, Billy Sol Estes. Bill Sol was on trial for fraud involving fictitious or empty ammonia tanks, and there were other suspected misdeeds including murder. He was well connected politically in Texas—especially with Lyndon Johnson. The night before they were to announce the verdict, you could have gotten a thousand to one odds at the Red Barn cocktail party that he would not be convicted or would be pardoned thanks to LBJ's influence. Apparently it was too hot for LBJ to handle because Billy Sol ended up getting a long sentence in a Federal prison. Some of these prisons are not the country club some might think. A friend called me from one of those prisons, and when I asked him what it was like he told me, "Doc, it's not worth a damn. There's a bunch of crooks here." The El Paso trip was fun and exciting, and thereafter I was hooked on travel.

My favorite hotel in the United States is probably the Madonna Inn in San Luis Obispo, California. Alex Madonna, Caroline and I were good friends. When we built two new houses on the California Ranch, Alex wanted to decorate the bathrooms, but we didn't think we could afford it. He did make the wedding cake for our Ranch wedding, and true to Alex, it was pink and a banana cake. The Madonna Inn rooms were great, the food was great, and Alex and his wife were wonderful hosts.

Without doubt, my wildest hotel story took place at the old Roswell Inn in Roswell, New Mexico. It's since been torn down, but at one time it was the crossroads of eastern New Mexico. I checked in the first evening and headed for the bar. Ginny, the lady running the bar and I visited a while then she said, "I've got a couple of customers I need to take care of." There was a young fellow and a girl (both on the smaller side), and he was obviously hustling for her and selling her "working girl tricks." Ginny told them to leave. The girl left, but the guy wouldn't leave. She called the police. When they came in, he ran out. The police found the girl in a motel room, and arrested her but couldn't find her mate. I had gone to bed, but later the fellow came back to the bar looking for his girl and he was mad as hell. They chased him out again, but on the way out he set fire to a big pad at the front desk that contained the wake up calls for all the rooms. He got away again, but they had to reconstruct the wake up calls for everyone in the hotel. They knew a lot of their guest's habits, but I was new. About midnight they called me to ask what time I wanted to wake up. I told them, "Damn sure not at midnight!" About 2 AM Ginny called and said, "I'm going to bring a drink over because it's been a hell of a night." After they closed the bar, she was got into her car in the parking lot. The fellow they had been chasing was in her back seat and started to strangle her. She yelled and the cops came and arrested him, but he put up a hell of a fight—but that's not the end of the story. Ginny served as special deputy because she was up late, and if they had a female they needed to check into jail, they would call her. They called and said "Ginny, you better come down here. We think this guy we arrested is a girl." She did and he was. He might have been a female impersonator, and small but apparently "it" was tough. Ginny and I became pretty good friends because she was also a part-time western wear model and traveled a lot. She was a country girl at heart, and could be exciting but she died at an early age.

As far as airplanes are concerned, I've had two or three private planes, but never flew myself. I figured I had the wrong attitude and mentality. I've always said I don't care about anything that doesn't grow hide or hair and needs maintenance. Obviously that makes me poor pilot material, but I had two good pilots in my career. The first pilot was L.D. Beedle from the Plainview, Texas area, and the second pilot was Scott Schroner a country boy from the Elkhart, Kansas area. They were both good pilots and loyal employees, and while I don't miss my planes, I do miss them.

My first plane was a low time, 1975 V Tail Bonanza I bought from Rex Reeves, the Clayton, New Mexico banker. Rex said it was a good plane, but he needed all the money he could raise for cattle loans. The Bonanza and I got along fine. Around 1987, I tripled up on feedlot numbers and made quite a bit of money. I used it to buy a couple of ranches (more about that later), and had not yet reduced my consulting. Friends began to tell me how busy and smart I was, and I really needed a bigger faster plane. What they meant was they wanted a bigger faster plane to fly around with me. It's nice to have friends brag on you, but it's a mistake to believe them. Unfortunately I did and I bought a pressurized turbo P Barron. It was a high maintenance nightmare. About five years later I sold it, and I've always said two of the happiest days of my life were when I sold my last airplane and my forest lease.

I loved flying commercial when fares were low, stewardesses were friendly and there were no security problems. The first time I went through security and had my briefcase examined was after some hijackings to Cuba. I connected from Shreveport, Louisiana through Dallas Love Field. I was working with some cattle people in the Shreveport area. They were showing me their new crawfish ponds and bragging about their great business. I don't know about you, but eating crawfish has always been more work than it's worth to me. I made fun of their crawfish project, and to get even, they slipped a bunch of crawfish in my briefcase before I got on the plane in Shreveport. When going through security at Love Field, they said, "Can we examine your briefcase?" and I answered, "Why not?" Out jumped about a dozen crawfish and the lady inspector screamed, and the cops came running up and saying, "How do you explain this?" I told them it was an example of having some bad friends.

Shortly after the 747s started flying, I got on one in Chicago going to Phoenix. I flew first class and it was great. However, after we had just boarded there was a bomb threat so we all had to deplane and hang around the airport bar for a couple of hours. We re-boarded, then about an hour out of Chicago we were going through some clouds and turbulence and there was a loud "crack." I looked out on the wing and there was a large ball of fire in the middle of it. Also, it knocked out all the communications in the plane so the pilot and the stewardesses couldn't communicate. It was interesting because everyone was frozen in place. About a minute later the pilot came on informing us that the plane had struck by lightning. When I saw that ball of fire, I first thought, "Damn, there's that bomb!" and my other two thoughts were interesting but predictable. "It's been a good life," and "At least I'm going out first class."

Another scare came one night on a flight from Denver to Grand Junction. After we were airborne, the pilots came on the intercom notifying us that a warning light indicated the nose gear had not properly retracted. Rather than going to Grand Junction, it would be easier to repair and they had more emergency equipment in Denver. They proceeded to circle for two hours while they emptied the fuel tanks. During those days, when you had this type of event they would serve free booze to whomever wanted it. Most of us did. About halfway through our circling, a group of Spanish nuns began singing religious songs in Spanish, and that made people more nervous. A fellow and I were sitting on the exit row. I was on the aisle and he was near the exit door. When we got ready to land the stewardess came by telling us how to get down into the crash position, and if we do actually crash when we come to a stop, pull the door open and get out and let everyone out behind you. The fellow next to me by the exit door said as we landing, "What I do is pull the door open as soon as we hit the runway." I said, "No, you wait until we crash or stop." He started arguing with me, and I slapped him and told him to get his head down and not look up until the plane stopped. Had he done as he planned, he and I probably would have been sucked out on the runway and everybody else would have been fine.

My most embarrassing incident was flying from Amarillo to Phoenix one Saturday morning on TWA. There was only one other person in the first class section, and she was a strange-looking young lady. In all of their wisdom, they gave us seat assignments next to each other. I was tired and just after takeoff, dozed off when all the sudden she started screaming. This was the first flight I have been on where rather than stewardesses they had two male stewards. One male steward came running up, and she said, "This man is molesting me." I told him this was a bunch of crap, and I'm gonna take the back first class row, and please keep the woman away from me. I obviously couldn't go back to sleep. The steward came back up and started talking to her, and she started screaming again. The other steward came running up and she said his partner was molesting her. That made me happy because it took some of the pressure off me. I'm sure the stewards relayed the problem to the cockpit, and soon the woman got up and started pounding on the cabin door yelling that both stewards were giving her trouble. The co-pilot put her in the front corner seat, strapped her in, and said if you make one more move, we're going to land in Albuquerque and put you off the plane. She shut up. When we arrived in Phoenix, the police came aboard and escorted her off. It ended okay, but I can tell you it is real lonesome at 30,000 feet when a woman you've never seen before and never touched accuses you of groping her.

One of the things that makes travel easy and relaxing for me is that I can be asleep before the plane gets off the runway if I don't have anything I particularly want to do. We had a free companion pass on Southwest for over twenty years because I flew with them so much, and Caroline always went with me. Annie travels with me now. If I didn't go to sleep, we'd get a lot of work done on the plane.

The nice thing about traveling before all the security problems is that you could have meetings in the airport at one of the hospitality suites. You'd get more done there than you would in an office. I would sometimes fly from one end of the country to the other and would hold important business meetings in airports two or three days in a row. If I was really busy and could make the right schedules, I would fly all night, clean up in the airport hospitality room bathroom and never go to a motel. I believe my record is three days of air travel without a motel room. My biggest problem in those days is that they lost my bags frequently, and my bags wouldn't catch up with me. As a result I'd buy pieces of a wardrobe at airport shopping malls, and needless to say, my wardrobe wasn't well-coordinated.

I guess one of my greatest aggravations about the airline industry and government is how they've screwed up the service and employee manners of a great industry. They no longer seem to comprehend or care. The exception is Southwest Airlines and if I can't fly Southwest, I'm likely not to fly.

The best example of how flying has become more complicated is that years ago we owned part of a travel agency. I'd have tickets written up ahead of time for different destinations. If I was flying privately, and near a commercial airport where I had a ticket to fly out, we would sometimes taxi right up to the commercial plane. At that time we boarded outside on air stairs. I'd get my bags out of my plane along with my ticket and walk up the stairs and onto the plane. Try that today and they not only arrest you, but probably confiscate your plane. I've given some thought to buying another plane but life is less complicated without one.

Just a Rolling Stone

K. S. Eng

Each of us has met them
Men that are born to roam
But we never really know them
Because they're like a rolling stone

Some of them are cowboys
Proud of the horses they've rode
Some of them are miners
Still searching for the Mother Lode

Not always quiet or from the country
They could be salesmen talking loud
You may find them in an airport
Lonely faces in the crowd

Each of them has a story
Of things that might have been
Many of them are lonely
Far from friends and next of kin

It seems they're always restless
They're men who can't stand still
They search for gold at rainbows end
That's always over the next hill

They mourn their youth that's lost
Even though they're past their prime
They still look for the exotic and new
Though they are running short of time

Late night they're sometimes lonely
And search for a port in the storm
They may look for it in a bottle
Or a body that's soft and warm

Next morning they'll start again
After coffee to clear their brain
Trying to recoup yesterday's losses
Hoping today won't be the same

Who knows what may drive them
Maybe it's the gypsy in their soul
Or a story of a loved one lost
Causing sorrow and their woe

So wherever you may meet him
Remember he's likely far from home
Take time and listen to his story
He may be a lonely rolling stone

The Curse of the Curious Mind

K. S. Eng

He tried to be considerate
He never meant to be unkind
But it seems the devil gave him
The curse of the curious mind

When all was going his way
He might leave success behind
And begin a new adventure
To satisfy his curious mind

He was willing to risk it all
Even if it meant a financial bind
It was the thrill of the unknown
Plus the curse of the curious mind

He was always ready to travel
Where ever the trail would wind
Had to know what was over the hill
It was that cursed curious mind

He was the ultimate risk taker
Fear of losing had no sting
He'd rather back a long shot
Instead of betting on a sure thing

Even when love and life was good
It seems he often pined
For a change in geography
It's the curse of his curious mind

When the final bell is rung
He prays at last he'll find
The Lord will have mercy
And satisfy his curious mind

The Drifter

K. S. Eng

Don't know what made him a drifter
But I think it just might be
Though his blood has genes
of a Swede
His heart had the soul of a gypsy

He traveled from ranch to ranch
They say he was a "hell of a" hand
But he was always moving on
Searching for new faces and new land

With good manners and a sly smile
He was a hit with the fairer sex
But when they tried to settle him down
They moved from his steady to his "ex"

He understood cattle and horses
Or maybe it was the other way around
Just when you thought he'd
found a home
He would disappear without a sound

I met him again at a Nursing Home
Visiting an old friend waiting to die
He said "these folks don't ask for much
They're just happy when I stop by"

He said "Could I ask you a favor"
His face was tired but still had pride
"When you need a hand for day work
I've still got a good horse to ride"

-CHAPTER 20-

Writing—Lots of It

For reasons unknown, writing has always been easy for me. In my lifetime, I've written approximately 600 technical and popular articles and poems. Some may suspect this is due to a big ego, but my ego was satisfied a long time ago. However, it could be easily be due to other flaws. Most of my article writing is due to requests or because it was a way to collect and focus my thoughts on a particular area. An unexpected talent for writing can have its downside. As a student I had two major college papers kicked back because they didn't think I was capable of writing them.

Writing the *Feedstuffs Bottom Line* articles was fun and fulfilling years ago when *Feedstuffs* was the premier publication in the industry. Responses to some of my *Feedstuffs* articles were entertaining. One of my first articles concerned the nutritional wisdom of sheep to select the correct free choice mineral supplement. The bottom line was sheep appeared to have very little inherent nutritional wisdom. The work was published by the University of Minnesota. One person who wrote said he didn't believe the research because the University was a bunch of 'pinko liberals." Another from a mineral company said I might be right, but to question nutritional wisdom was like questioning Mom and apple pie. He predicted I wouldn't last long as a *Feedstuffs* writer.

I wrote an article about an Oregon State study comparing the intelligence of various animals and birds. Of course chickens ranked tenth out of ten. A lady who apparently loved chickens wrote me a four-page, single-spaced letter describing how smart chickens really were and their problem was they had been raised in cages and mistreated by humans. She closed by saying that if I had been raised in a small cage and treated like a chicken, I might not act smart either. I agreed with her and that probably ruined her day.

After an early article on beta-agonists (1985), I received several inquiries from "show calf and veal producers" and some body builders. I once received a call from a government official asking if I knew of any unusual demand for Trenbolone implants. I replied that anyone in our industry might be considered unusual. He didn't think that was humorous and was worried about its illegal use by body builders. I told him not to worry because when it came time to re-implant, their ears would be sore and they would probably discontinue it. He didn't think that was funny either and informed me that the route of administration in humans was as a suppository. The conversation ended when I asked him if that was an Aggie joke, and he hung up.

After thirty years of writing the *Bottom Line* column, I felt it was past time to say *adios*. I received several calls asking me to continue, but I wouldn't reconsider, and besides *Feedstuffs* had a new strategy in place for the *Beef Cattle Bottom Line*. Several people who contacted me were unhappy with *Feedstuffs* and some with me. Occasionally I still receive requests and I suspect the reason is there once was a genuine affection for the old *Feedstuffs* when it was no doubt the dominant magazine in the industry. It proves once again few things last forever. The company that took over my *Feedstuffs* writing told me they didn't realize how much work it was and planned to quit.

On the positive side there are still a number of good Ag publications available and the quality of the writing is excellent. It appears several publications continue to thrive even in the era of wireless communications. Bottom line: the Ag publication industry appears to be doing fine without me and that's how it should be.

I have no explanation for my poetry writing that really didn't begin until approximately fifteen years ago. I had written a few poems prior to that and occasionally I would see them show up uncited in publications so I decided to put out a small book of poetry and copyright it. That was around Y2K, and once I got started, my poetry writing became spontaneous and prolific. Most of the poems, some of which are scattered throughout this book, were written in five to ten minutes. Often Caroline was my inspiration and muse. Almost all are based on personal experiences or experiences of friends. It was as if somebody would tap me on the shoulder and say, "Write this down and publish it."

An example of poetry writing that was spontaneous and based on a true event is "Working Girl" found on the next page. One night Tom Peters and I met late at a bar in the Del Coronado Island Hotel. He was sitting next to an attractive girl. Tom told her, "Hey this is Dr. Kenneth Eng who is a famous poet. I'll bet he'll write a poem for you." With that, Tom left. I couldn't come up with a poem, but she and I visited. She was obviously a pretty high-class working girl. About two years later, I woke up in the middle of the night, grabbed my trusty pen and yellow tablet and wrote, the "Working Girl" poem in less than five minutes. When Caroline typed it the next morning she asked what the inspiration was. I blamed it on Tom Peters, and regardless, it's still one of my favorites.

The Working Girl

K. S. Eng

You knew she was a "working girl"
There were no calluses on her hands
But her eyes and body looked hard
From too many "one-night stands"

She sat next to me at the hotel bar
Asked if she could buy me a drink
I said "I'll just sip my beer
I need a clear head to think"

"Think about what she asked"
I said "money, my ranch and cattle
I've got a note that's long past due
And I'm mighty short on collateral"

She said "my husband and
I had a ranch
It was a small one far from the city
When the drought and bad
markets got us
The creditors had no patience or pity"

"We split up and I moved to the city
Far from my country roots
Now my work dress is a low cut gown
But I still miss my jeans and boots"

She said "Cowboy here's
wishing you luck"
Tipping her drink and her low cut gown
"God speed getting back to your ranch
You don't fit in this cold hard town"

"And when you get back to the ranch
If you ever need a hard working hand
I'll trade my gowns for boots and jeans
And a chance to get back to the land"

I wrote the poem "As Long as I Have You" in three minutes while riding with Caroline. She brought out the best in me. My poetry writing pretty much came to an end when Caroline died. Most of my recent poetry is too gloomy and melancholy. Notwithstanding this book, I suspect my writing is near its end. I still receive requests asking I write more, however, I believe that's unnecessary because there's an abundance of good younger writers and good publications.

As Long As I Have You

K. S. Eng

Late at night I sometimes wonder
What am I going to do?
When my youth and vigor are gone
Will I be lonely and blue?

Ours isn't the biggest outfit
I'm not the leader of the band
But we've made our own decisions
And owned our cattle and land

Lord I love the freedom
Though we didn't get all the breaks
But even when times were tough
We could make our own mistakes

So if we have to move to town
I hope there's some chores to do
Than I know that I'll get by
As long as I have you

Bud McCaulley—My Artist and Friend passed January 3, 2014

Bud McCaulley

Bud McCaulley, age 81, passed Friday, January 3rd, 2014.

No services are scheduled. The family welcomes friends to call and come by for a visit. Arrangements are by Boxwell Brothers Funeral Directors, 2800 Paramount Blvd.

Bud was a veteran in the United States Marines. He was a successful artist in Commercial art, fine art, and Cowboy cartooning.

Bud will be missed by his wife Gwen, daughters Cathy, Chris, Susan, Jan and son-in-law Sam.

Bud and Gwen have been blessed with eight grand children and five great-grandchildren.

Sign the online guest book at **www.boxwellbrothers.com**.

Bud McCaulley:
1932-2014

Amarillo Globe-News, Jan. 6, 2014

Those of you familiar with my poetry books and calendars are aware of the great cartoon artwork Bud McCaulley contributed. However, there was much more to his life and career. He was born in Kansas City but had resided in Amarillo, Texas the past fifty years. He was well known for both his cowboy cartoons and his serious art. His series of western cartoons known as Bluebird prints are classics. McCaulley Did It prints and cards are found in western stores and on coffee tables throughout America. In addition, he was commissioned by the State of Texas to create portraits of forty-six Texas Presidents and Governors to celebrate the Texas Sesquicentennial Anniversary in 1986. This collection hangs in the

State Capitol Rotunda in Austin, Texas. He was a founding member of Cowboy Cartoonists International (CCI) an organization dedicated to upgrading the image of the western cartoonist.

He was a veteran of the United States Marines, sang in barbershop quartets, traveled and fished. His career as a fine artist is lesser known, but some of his paintings are outstanding. I have an original he painted of a mountain lion that is the best wildlife painting I have ever seen. He is survived by his long-time wife, Gwen, and several daughters and grandchildren.

Most of all, he was a good friend as well as an entertaining fellow to hang out with—in short, a good human being. Along with numerous friends, I will truly miss him.

For several years there were many good artists in the Amarillo area. One of my other favorites was "Peaches" Angel Stevens. I have several of her paintings in my collection. She never made much money painting, but she had a creative mind and personality. Her other endeavors included teaching microwave cooking, reading palms, massage therapy and hypnotism. She was never boring—a free spirit who died too young.

"Peaches" Angel Stevens

-CHAPTER 21-

The Decade of the '80s (Lots of Industry and Personal Changes)

Changes in interest rates, tax laws and the banking industry began in 1980 and continued throughout the decade. In 1980, prime interest rates reached twenty to twenty-two percent which impacted our borrowing ability, our non-feed costs and break-evens. Beginning in the early '80s, the governments began to question prepaid feed and tax shelter cattle investments, and by 1984 many of these programs were disallowed. This meant less participation by outside investors in cattle feeding business. Some welcomed this turn events, believing it would decrease feeder cattle demand and prices. Obviously, long-term this proved incorrect. We've demonstrated that the cattle industry, without outside help, can push calf and yearling prices to all-time highs. In the mid- to late-eighties, the banks themselves got into a lot of trouble—not because of livestock loans, but because of bad housing, real estate and development loans. There was a time when the cattle industry was at a greater risk of having their wagon hitched to a failing bank than vice a versa.

In general, I've been fortunate to have good banks and bankers. That's as it should be since we never lied to them, and always repaid our loans on time even though sometimes it was nip and tuck. However, during that period many legitimate cattlemen had unwarranted problems with their lenders. One bank I was involved with for a short period was a real pain in the butt. We even discovered that one bank that we did not owe any money showed liens on some of our cattle. Occasionally it's mind-boggling how ignorant and arrogant a few bankers can be.

From a technical standpoint, there were several implant and implant combinations introduced in the mid- to late-80s, and choices of implant programs expanded geometrically. One scientist told me that with light cattle that might be implanted as many as three times there were over 3,600 possible implant combinations. I don't doubt it. and I may have had a few clients that tried all of them. Some thought the trenbolone acetate implant might help promote more growth and final size from small-framed light heifers. I personally tried this on 300-pound small-framed heifers purchased in Louisiana. I gave them an estrogen-trenbolone acetate implant three times during the feeding period. When finished, they were a little larger, but they looked and sounded like bulls.

I brought a fat cattle buyer by the pen one day, and he said, "Kenneth, I can't buy them, they're too stagy."

I replied, "What do you mean? They're long-fed young heifers."

He made me an offer which I took immediately. I never had the nerve to ask what the carcasses looked like.

The feed additive Bovatec was cleared in 1982. Roche developed the ionophore positioned to compete with Rumensin. I had done research on it, and was also on their research review committee that met quarterly in New Jersey. I believe our committee was made up of Jim Nofsiger, Dale Furr, Gene Erwin, Earl Bartley and myself. The research performance data looked outstanding, and Dr. Bartley from Kansas State, who was an acidosis expert, believed it was also extremely effective at preventing acidosis.

At one of our New Jersey meetings, Gene Erwin and I had just completed separate bovatec trials. As I recall, both trials showed a 4½ percent performance improvement compared to Rumensin. However, they began the meeting by announcing the FDA had just cleared the product, and they didn't want any more research information for the FDA to consider. Thus, neither the Erwin nor the Eng data was published. However, when they began mass production and marketing of the new product, they encountered a lot of metabolic problems and some questionable performance. There are several individuals who believe (I guess I am one of them) that the final product marketed was not the same as the experimental product we tested.

Milo, wheat and barley were better grain buys than corn in many areas of the country. I used less corn compared to other grains, whereas today many nutritionists have no experience with grains other than corn. Feed prices declined in the early to mid-'80s, especially some of the by-product prices. If you were alert and flexible, you could formulate some inexpensive rations with low-gain costs and low-projected break-evens. However, one of the biggest wrecks in feedlot history lurked just around the corner. As before, the government was directly involved. In 1986, the dairy buyout program went into effect whereby dairymen were subsidized to sell (that is, to give away) their cows and exit the business. Many of the dairymen made no attempt to orderly market their cows. The result was a sudden and severe price drop. I believe over 1.5 million dairy animals were marketed during the next eighteen months at a cost of 1.8 billion to the taxpayers.

Fortunately, I didn't have many cattle on feed at that time, but reasoned that the buyout program was so ill-conceived and stupid that the government would amend it. I placed every three-weight calf I could afford in Clayton Feedlots and when they were fat later in '87, fat cattle prices had rebounded. I believe my average cost in these calves was around $100 per head and I made enough money to go ranch shopping in 1988.

Farm and ranch land prices stayed under a lot of pressure in the 1980s and bottomed in '87 and '88. I began scouting for a large ranch in either Nebraska or New Mexico and was soon covered with offers. Some old-time cattle traders accumulated both cattle and ranches based on the assumption that one of the markets would be always be good. Several of them went broke when both the cattle and land markets collapsed. When noted border trader Jeff Berner died in 1987, his widow and stepson, Butch Stevens, asked me to look at the ranches he held in New Mexico, Colorado, Wyoming and South Dakota. They were upfront and said, "If we don't sell, we'll lose them, and we'd rather a cattleman get a good deal than the bank." We looked at several good ranches—most priced around $70-plus per deeded acre. One near Spearfish, South Dakota that was ten thousand acres was priced at $35 per acre. Also Clayton banker Rex Reeves had found a ranch in Sierra County, New Mexico that was either going to sell or go into bankruptcy. In addition, I looked at a large ranch on the Loup River near Brewster, Nebraska priced at $65 per acre. At this time, Ed Cox Jr. still owned the majority of our California ranch that I wanted to buy. To sum it up, I was like a kid in a candy store and spent months looking at ranch properties and finally and bought two.

In the mid-'80s I had taken on some Asian clients in Japan, the island of Hokkiado, Indonesia and Hong Kong. I really enjoyed it and met some very gracious people. Seiji Terada was my Japanese interpreter. He brought several groups back to the Southwest United States to attend meetings I conducted, and also to visit some clients and my ranches. These were good times and good people.

A pleasant and exciting new interlude in my life began in early 1980. I was in the Phoenix airport and a lady named Sibylle (Billie) and her mother who was from the Black Forest area of Germany inquired where the gate for TWA Albuquerque could be found. I showed them, and since they weren't leaving for a couple of hours, I took them to the TWA hospitality room. I mentioned I was flying to Albuquerque the next morning, and my pilot was going to pick me up. I told Billie if she wasn't doing anything special, to meet me at the airport and I'd show her some new country. I assumed I would never see her again, but she was waiting at my gate the next morning. It was beginning of an exciting seven-year venture.

Billie was smart, fun and adventurous. She would come to the U.S. and stay three to four months until her visa ran out, and then she returned to Europe. Occasionally I'd travel to Europe or London to see her or to the Black Forest area that is beautiful.

Billie liked the beach better than the dusty trails.

All went well until she decided to move permanently to the U.S. She asked what I thought about La Jolla, California. We visited the area and I said, "It looks okay," and I bought a house there. She gradually lost interest in traveling, and I lost interest in La Jolla. Most people would love living in La Jolla, but it wasn't for me. I always said that once you finish discussing wine, cheese, your horoscope and your tennis game the conversation in La Jolla was over. I'm not very good at any of those things. We had six great years followed by one aggravating year. Six out of seven is not bad. We remain good friends, and she calls occasionally. Billie speaks several languages but English isn't her best. She called once to say her brother in-law in Germany who's also a friend of mine had died suddenly. I asked her what he died from and she said "hemorrhoids." I said, "Billy, hemorrhoids can be painful but I don't think they'll kill you. She replied they will if they're on your brain. I said, "I think you mean hemorrhage," and we had a good laugh and agreed that hemorrhoids on the brain would be painful.

The late '80s were a busy time for me from a business and personal standpoint. Like many in the cattle business, the dairy buyout cost me "big bucks," but I was able to recoup. In the late '80s, I "tripled up numbers" feeding light calves and profits were enough for a downpayment on a couple of ranches. More about this later.

Most important, out of nowhere Caroline called in the summer of 88. I hadn't spoken to her in almost eighteen years, and she soon became my soul mate, constant companion and working partner. She was the missing piece in the puzzle of my unique lifestyle. Once again, timing was everything and perfect timing is truly a gift from God. From 1988 until she drowned in 2010, we had our share of problems, but every week was like a week at the National Finals Rodeo. It was akin to drawing a tough bull or bronco, and you knew if you could hang on, you would finish in the money.

Covering Losses (Plead insanity)

K. S. Eng

Guess the banker was worried
He decided to check around
The cows mortgaged in Oklahoma
Were nowhere to be found

Next he traveled down to Texas
To check the yearlings out on wheat
He found no steers and the wheat
plowed under
Then turned pale and white as a sheet

They've locked me in the county jail
But I've lawyer who's clever and sly
He said, "Explain what you've done
And tell me the reasons why"

"Well it started with those Mexican cattle
I bought from a trader of some renown
I paid for the calves in Juarez
In a cantina in old Boys Town"

"The trader and the cattle vanished
Of course I had to cover my loss
Went to Chicago and traded pork bellies
And they nailed me to the cross"

"Then I borrowed and bought
race horses
Invested in some cutting horses too
I ended up buying used cows
What else could a trading man do?"

My lawyer listened carefully
With a look that was pained, but kind
He said "you should plead insanity
You're obviously out of your mind."

New Loan Strategy

K. S. Eng

When you go to the bank for a new loan
Ask for twice as much as you need
Speak only to the President
Show your arrogance and your greed

They prefer to deal with the big guys
The little fellow hasn't much of a chance
Dicker on the interest rates
You're not in business for the romance

Brag on your outside investments
They're whatever you'd like them to be
For example those
short-term activities
Could be your tickets to the lottery

Talk of your investment seminars
They'll know you're a pro, not a rookie
They needn't know your
financial advisors
Are a bartender and your bookie

Whisper of foreign investments
That kind of intrigue they love
Don't tell them they consist of
Jose Cuervo and Canadian Club

Act like you don't need the money
Tell them you've plenty of your own
Make believe you're doing them a favor
So they think you'll pay back the loan

–K–

Too Much Barbeque

K. S. Eng

"I'd prefer a window seat"
She told me with a smile
I said, "This is serendipity
Because I prefer the isle"

As we waited for our baggage
She said, "I'm a stranger in this town"
I said, "Why don't you come with me"
I'll be glad to show you around

I rented a four-wheel drive
As we were driving from the city
I thought this is my lucky day
Cause lord, the lady's pretty

We drove 100 miles south
I checked some Mexican steers
On the way we made a pit stop
For barbeque, beans & beers

Next day we drove to a feedlot
Looked at cattle & talked to the boss
Than we stopped for brisket barbeque
Where they serve my favorite sauce

That afternoon we went to an auction
I bought enough calves for a load
Then drove to a family barbeque
About 30 miles down the road

Next morning we checked a cow herd
She said "let's try a new menu tonight"
I replied, " I know a great place
They do barbeque goat just right"

"I've enjoyed all I can stand", she said
Though she said it with a smile
She flew East by the window
And I flew west on the isle

I had asked could we meet again
She said, "Sure, anyplace would do
Just so it's 1000 miles from a cow
And twice as far to a barbeque"

-CHAPTER 22-

Ranching in New Mexico

The ranch I bought in New Mexico was approximately 70,000 acres located in Sierra County next to the Gila National Forest. It bordered two small towns—one was Winston with a population of about forty, and the other was Chloride which was an old silver mining ghost town with a population of about twenty-five. As the crow flies, it was about 170 miles southwest of Albuquerque or 170 northwest of El Paso and approximately forty miles west of Truth or Consequences (T or C). As you may know, T or C got its name from the TV show by that name which was hosted by Ralph Edwards. Ralph Edwards offered to do a show in any city in the United States that would change their name to Truth or Consequences. They did and he did. In fact, he continued to visit the area every year for another forty years until his death. The forest lease was estimated at 36,000 acres, although no one was certain. As an indication of how rough it was, the cow permit was for 243 head (4 cows/section). After the lions and bears got their share of the calves, they could have given it to me, and I would probably have lost money on the cows.

The main industry in the area was ranching and hunting elk, mule deer, lions and bears. At one time there was a pretty fair timber industry in the area, but the environmentalists took care of that. Additionally, there were good silver deposits and some copper. The rest was ranching which was a combination of cows and yearlings.

Sterling and Judi Carter who ranched nearby had a guide service. They managed my ranch for the first couple of years. I learned a lot from them, but like most newcomers, I also had to learn from mistakes of my own. I wanted to run cows and took a chance buying a herd of young Beefmaster cows from South Mississippi. Had they been Beefmaster cows from New Mexico, I'd have been okay, but these cows were big and fleshy and probably averaged 1,300 pounds. My ranch couldn't support a cow that large. They were bred when I bought them and weaned good heavy calves. Unfortunately, that is the end of the good news because the next pregnancy check about forty percent of them were "open" (not bred). That trend continued until I pretty much liquidated my big Beefmasters and replaced them with smaller, uglier desert cows.

Our Gila Forest lease was rough and I had no desire to stock it with cows at the rate of four per acre. The best deal we had was the first two years when Sterling helped me convince the Forest Ranger that on a 243 cow permit, we should be able to increase the numbers if we put out stocker cattle and we only left them out four months of the year. Then we were be able to run fifteen hundred head, and it worked well until that ranger retired. They then informed me that I had a cow permit, and if I wanted to run yearlings, I needed a new permit that would take at least two or three years to obtain. It had quit raining in the area, and I took three continuous years of non-use on the Forest permit and then sold it. Now you know why I've always said that other than getting together with Caroline, the two happiest days of my life were when I sold my last airplane and when I sold my Forest permit.

We had a couple of extra houses on the ranch allowing my pilot Scott and his family to move there and to help manage the place. We had a good airstrip on the top of a mesa not far from the main house, but it had mountains on two sides and a deep valley on each end. Take off and landing could be thrilling. Like our California airstrip, we also found out that airstrips in isolated areas can be attractive to drug traffickers who are good for people to avoid.

I redecorated the main house that was halfway between Winston and Chloride and moved there in 1988. A footnote here: I retained an interior decorator that was good, attractive and friendly. My advice to anyone who wants to keep their interior decorating budget under control is to hire a less attractive decorator. Regardless, I ended up with a nice house and office. I lived there part-time and the rest of the time in either Texas or California.

Perhaps, the most fascinating thing about the New Mexico ranch were our neighbors and the people in general. They were quite a mixture of ranchers, miners, outfitters and hunters and the like. Some obviously had moved to the area to get lost. One of my favorites was a fellow we called Indian Johnny who claimed to be a full-blooded Apache and I believe a relative of Geronimo or Cochise. I found him sleeping in an old Blazer on a remote part of our ranch. He asked if we had any work. It just so happened we were doing some fencing so I said, "Sure." He was a fair hand. He also had a great imagination, plus a fixation with helicopters and firetrucks. Shortly after we hired him he wanted to know the ranch mailing address. Before long we received a big package addressed to him as Assistant U.S. Fire Marshal. It contained badges and other items, and for a while, he was with the Volunteer Fire Department. He would often walk around in a slicker and a fireman's helmet. At a Fourth of July party, he asked me what kind of a doctor I was. He said he was looking for a plastic surgeon. When I asked him, "Why?" he said because he wanted his real face back. He claimed to have been in a firefight in helicopters, and his helicopter crashed. He was taken to the hospital where they redid his face so he couldn't be identified.

I asked, "Was this with the CIA down in Latin America?"

"No," he said, "It's deeper than the CIA and it was over the New Mexico-Arizona border."

Another colorful local looked like a 1960s refugee from the Haight-Ashbury San Francisco area. He asked for work one time, and we told him he could help on the yard. I had a lady named Sharlene working for me part-time who chose to direct his efforts. He came in a day later and told me he resigned because there was just too much pressure. His girlfriend who had also worked in the yard part-time managed to steal one of my credit cards. Before she was done, she had charged a lot of money. I pressed charges against her, and they put her in the T or C jail. Her hippie boyfriend came to see me and asked if I would drop the charges if he paid the bill. I said, "No, I can't do that because I don't want the word to get out that I'm an easy mark in the area. Besides I noticed she was keeping some pretty bad company." He indicated he had also noticed, and that was the end of the conversation.

The lady in jail did call my house at 4 AM one morning. I wasn't there, but she woke up Sharlene. The jailbird asked if I would drop charges.

Sharlene asked, "Why are you calling at four in the morning?"

She said, "That's the only time I can get the phone."

After she hung up, Sharlene called the County Sheriff.

The Sheriff asked, "What do you have on your mind?"

She informed him, "People in your jail are calling at 4 AM, and each time they do it, I'm going to call you back." That was the end of that problem.

One other humorous hippie boyfriend incident happened during a hard rainstorm. I saw him standing on our porch by our back door. I opened it and asked him to come in. He said, "I'm too wet, but it's raining so hard I can't see out of my van." I asked if his windshield wipers weren't working and he replied, "I don't have a windshield."

In their own way, the people in the area were pretty smart, fit the area well and were good entertainment. There were several other good people in the area and I learned more from them than they did from me.

The one thing that drove me crazy on the New Mexico ranch is that there were a lot of roads crisscrossing the ranch plus two gravel state highways. Tourists and visitors were always driving off-road across pastures and wherever. It was not unusual to find someone stuck in a canyon on private property where they shouldn't have been. It was a pretty area and had quite a few Indian artifacts that drew outsiders. Because of all the roads and the traffic, I decided to start selling parcels if people wanted to see it so badly. We ended up selling over 12,000 acres of land in parcels ranging from forty to eighty acres and sometimes 320 acres. Caroline took care of this project and drew up the contracts. The standard contract was 25% down with ten years to pay at 1½ to 2 percent over prime interest. We never ran an ad, and we only had one bad debt. Caroline—ever the banker—was good at that sort of thing, and she loved to handle money. Furthermore, when I met her in 1988, she had barely survived the Texas banking disasters. Two banks where she had been an officer and stockholder had been bought out, but then

Caroline and a Brahma on the New Mexico Ranch

were later closed. As a result, she was essentially broke which she wasn't accustomed to and damn sure didn't like.

It's funny how things that are either minor or major irritations sometime work out for the best. Had it not been for the problem of maintaining privacy on the New Mexico ranch, I might not have split it up. After Scott left, I hired a Texas cowboy and his wife with the idea he might control some of the traffic. Hank was a good cowboy, and I really liked him, but it was evident that he had a "really hard edge." He got in a fight or two and did well, but I didn't want him roughing up some tourist and ending up in court. In the end, he went back to south Texas where unfortunately, he was shot and killed.

To end the conversation on the New Mexico ranch on a humorous note, my ex-wife, Pat, called me shortly after I bought it and said, "I heard you bought a big ranch in New Mexico and I suppose you did it just to aggravate me."

I asked, "Why do you say that?"

She said, "You know I like New Mexico, and want to move there someday."

I said, "Well, it's a big state and it's an interesting state." I went on to tell her that there were more Ph.D.'s and a higher rate of illiteracy per 1000 in New Mexico than in any other state in the Union.

Her reply was, "That's because they count some of you in both categories."

I told her, "If you had been that funny when we were married, we'd have probably never divorced."

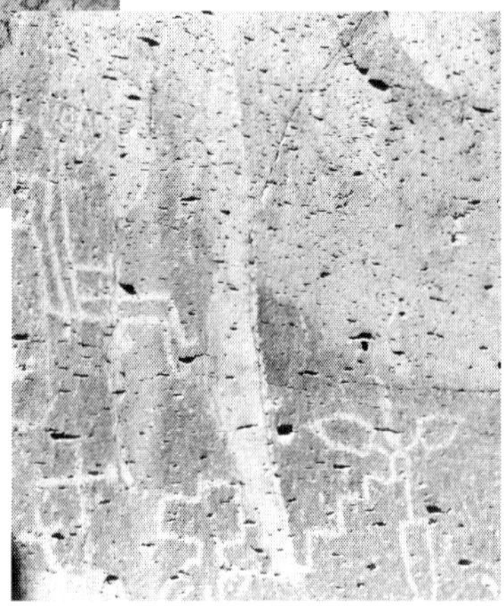

Native American Petroglyphs found on the New Mexico Ranch thought by archeologists to originate with the Celtic culture 2000 years ago.

Hank and I after turkey hunting.

Engs and Parkers bear hunting on the New Mexico Ranch.

End of the Trail

K. S. Eng

They grew up as childhood
sweethearts
And their love would never fail
They searched for gold at the
rainbows end
Hoped to find it at the end of the trial

They always yearned for
something better
Never satisfied with what they found
Didn't care where the trail led them
As long as it wasn't close to town

His ancestors were early Trail drivers
There was Indian in her veins
They hailed from the land
of enchantment
Ranching and cattle were their games

When younger they broke horses
Till their bones became too brittle
They were always short of money
But they needed very little

Mostly they worked for other ranchers
Sometimes they ran cattle of their own
But times changed and outfits
got bigger
& they were too small to get a loan

Recently they came looking for work
They were driving a battered truck
I said sorry, I couldn't use them
Fed them and wished them luck

Today I just learned what happened
In the Livestock Weekly I read the tale
They missed a curve on a
mountain highway
Together, they reached the end
of the trail

-CHAPTER 23-

Buying the California Carrizo Ranch (A Dream Come True)

In 1971, Ed Cox, Jr., who was a new client, and I flew around the country in his Sabreliner jet looking for a place to run winter grass cattle. I told him my choice would be the central coast of California and I had some contacts there. To make a long story short, Ken and Robert Palm picked us up at the Paso Robles, California airport and we drove out to the Carrizo and Chimineas ranches that they had leased. The Carrizo was my favorite, and in April that country is "dressed to kill." It was beautiful. Ken Palm, who was perhaps the best salesman I ever met, dropped the idea that we might possibly buy the ranch, and not just lease it. He had the lease and the inside track with the owner, which was a steamship company from San Francisco. It was a twenty-two thousand deeded-acre ranch.

Ed said, "What do you think we can buy it for?"

Ken said, "Probably $100 per acre, but you could try for a little less."

We decided to offer them $75 per acre and to go to a $100 if necessary. Ed wanted at least sixty percent of the ranch, and we could split the remaining forty percent. Ken went to San Francisco and offered the owners $75 an acre, and they accepted. Suddenly, Ed Cox Jr. had 60% ownership with Ken Palm, Robert Palm and Kenneth Eng each holding 12½ ownership of what I think is among the best ranches in the area. This was my first big purchase, and even though I was only a one-eighth owner, I was proud and happy.

We operated it ourselves for a couple of years running yearlings and dryland farming wheat or barley. I remember buying a Steiger tractor that cost $36,000—equivalent to three Lincolns at that time. We ran it 24/7, and sold it three years later for more money than it originally cost. Ed and I decided it was probably best that we get out of our partnership with the Palms and lease them the ranch. We were running yearlings and moved them back to Texas just in time. Shortly thereafter, Ken Palm filed bankruptcy. He had a large cattle loan with a Central Coast bank. When they gathered all their cattle after the drought and the wreck of '74 and '75, they were severely short on numbers. Ken went to prison at Lompoc. If you were going to prison, Lompoc is a good place because it's close to the coast on the southern part of the Central Coast area and beautiful country. He called me a few times from prison saying that he had an interesting roommate whom I believe was John Haldeman—one of Nixon's henchmen. Ken said, "I'm learning a lot from Haldeman," and I thought to myself, "I bet Haldeman is learning a lot too."

Crocker Bank held most of Ken Palm's notes. They took over Ken and Robert's share of the ranch. Ed and I told them, "Welcome to the group. We'll submit the bills for operation and expect you to pay them on time." Crocker bank never paid a bill. Finally we threatened to throw the ranch into bankruptcy, buy it back ourselves and eliminate them. They finally got out and that left Ed and I. I was still just a one-eighth undivided interest owner. To me, the sad news on the Palm brothers is they split, and seldom talked to each other again. Like all of us, they had their flaws, but they were involved in some of the best years of my life. At one time, the old Wilson's bar and restaurant in Paso Robles was the party spot of the Central Coast. People like Rambo, Palms, Sintons, Eades and others made for many memorable nights.

When Ed, the Palm brothers and I ran the ranch we kept two or three thousand yearlings plus we dryland farmed six to eight thousand acres. Its tough farming country, and we damn sure weren't the best farmers. I've proved over and over again that as a farmer, I'm a flop. Fortunately, I do know people who are good at it.

Rod Moss was Ed's right-hand man and we went to the ranch fairly frequently. After a couple of years it was usually just me. Because the Palms leased two adjacent ranches, at one time they kept a big crew on the Carrizo ranch with a bunkhouse and a cookhouse. A part-time cowgirl-cook did the cooking. She was a good hand at both, but "ragged around the edges." One lunchtime, Rod and I were eating in the mess hall, and both of us had big appetites. We decided to go back into the kitchen to see what was left to eat. As we walked past the sink, we saw it was full of dishes plus the cook's panties and bras. Rod and I lost our appetite.

After the Palms departed, Ed and I decided it would be best to lease the ranch out, and eventually the Rossi brothers from Bakersfield got the lease. They were big cotton and alfalfa farmers who knew how to operate a farm. They were also pretty good with cattle and Melvin's son, Mike, was involved. Unfortunately, Melvin died suddenly when he was around fifty years of age leaving all of their operations in the Bakersfield as well as our ranch for Mike to run. Mike did the best he could. He was a good friend. The Rossi's leased our ranch for approximately ten years. I bought out Ed and I took over the operation around 1990. In the meantime the Rossi's encountered financial problems which was not difficult for farmers in the '80s. When they left the ranch, they left three or four "Big Bud" custom-built tractors from Havre, Montana parked at our headquarters. They also left a lot of equipment which was custom built by Big Bud. In fact, they had the largest Big Bud ever built. I believe it had two Cummins diesel engines with a total of 760 horsepower. What I do remember distinctly is that the diesel tank on the tractor held over 800 gallons. It remained parked at our place for over four years. I would climb into the cab once in a while, and you could get a pretty good view of the whole ranch. Eventually, the creditors came and picked up the tractors, but I wouldn't let them have some other equipment because I was stilled owed money on the lease. It's not easy to find a market for a forty-foot disk and harrow, but I eventually did.

Buying the remainder of the Carrizo ranch was a real adventure. Although I had lost personal contact with Ed Cox, Jr., the entire industry was aware of his money and legal problems in 1986 and '87. My dream was to buy the rest of the ranch. When I contacted him I was turned over to an attorney who was your worst nightmare. She was both condescending and arrogant. I got nowhere with the buyout and finally gave up and bought the ranch in New Mexico instead.

Just after we had closed on the New Mexico ranch, Ed called saying, "Why don't you buy my portion of the Carrizo Ranch."

I answered, "Ed, I've been trying to for over a year and your attorney won't give me the time of day."

He said, "I bet we can settle this in a hurry. Where can we meet as soon as possible."

I told him, "I'll be in Amarillo tonight."

He said, "I'll meet you there."

Ed asked what I'd offer for his share of the ranch and I took a million and half dollars off of my original offer because of the attorney grief and because I'd already spent most of my money.

He coughed and said, "Let's go ahead and do the deal, but beware that when they do the title search, they're going to find some big liens against the property."

I asked, "How can that be, Ed? We have the ranch paid for."

He replied, "Don't ask questions, but that's the way it is. I'll get the liens cleared when we're ready to close."

I was able to get the financing from Farm Credit West in Paso Robles. When they were doing the title search, the lady in charge was like most title inspectors—kind of a pain in the butt. Her name was Paula and she was from Pismo Beach. We called her "Picky Paula from Pismo."

She called to tell me, "Kenneth, this deal never going to get done because there is almost eighty five million dollars worth of liens against the ranch."

The amount staggered me, but I told her do the paperwork, get it ready for closing and let's see what happens. When it came closing time, true to Ed's words, the liens disappeared and I owned the ranch of my dreams.

One other humorous part of that conversation was that Paula would call me almost every day to say, "I'm trying to call Ed Cox Jr. and all I can get is lawyers."

I said, "Paula, you've got as much chance of talking to Ed right now as you do Howard Hughes—just do your paperwork."

Although Ed and I have drifted apart, I'll always be thankful for him making the Carrizo ranch purchase possible.

It was widely known Ed encountered a lot of trouble in the mid- to late-'80s, and rumor has it he lost in excess of two hundred and fifty million dollars. Also, he ended up going to jail, but was paroled early. I've been told his conviction was wiped out by executive order from President George Bush, Sr. Obviously, Ed Cox, Jr. still had a lot of clout and likely still does. He was a good friend, and one of the smartest people I ever met.

Caroline truly loved the Carrizo ranch, and were she alive I probably wouldn't have sold the last of it. In the mid-'90s, I sold the original headquarters of the ranch to a neighbor, Sally Friend, and we built a new foreman's house plus the house of our dreams back in the mountains. Because of her banking background Caroline knew a lot about housing construction and we basically designed and subcontracted both of the houses in the mid-90s.

At that time the housing market and construction market was "on its butt." Thus we were able to get good help building the houses even though we were fifty miles from the nearest town. The main drawing point was each Friday when Caroline would write the workers a check. We finished our houses in record time. One minor problem on our main house is that I knew exactly how I wanted it to angle for the best view both morning and evenings. Unfortunately, the back of the house was on a rock outcropping that was a rattle snake den. Some of the rattlesnakes in that area are the real bad guys—Mojave greens. They have venom similar to a cobra, and they're both aggressive and deadly. Obviously, this excited the construction crew. The area tended to be rattlesnake heaven, and two years ago we killed over fifty rattlesnakes on the ranch. Mind you, we weren't deliberately hunting them, but they had the misfortune of getting in our way.

The other interesting part of the ranch is that we developed a large Tule Elk herd and were able to obtain Landowner Management Permits for elk hunts. If you're a true elk hunter, you know that the Tule elk is a unique sub-species that was once endangered and rumored to have only four left in the world. Fortunately, they were able to increase the herd in our area and marketed special elk hunts for several years.

The remainder of the eighties was relatively nondescript, but great for me personally. Cattle profits peaked in 1987, land prices bottomed in 1988 and most important of all, Caroline re-entered my life. Once again "timing is everything."

In the late 1980s, I started running quite a few yearlings in the southeast and on the New Mexico ranch. Many were in partnership with Harold Parker, and if you crave excitement, partner with the Parkers. In spite of tough times and some arrogant bankers, we did well.

Our biggest Tule elk kill.

One of my favorite places was Ronnie Thomas' place near Ocala, Florida. This is another short, sad but funny story. Ronnie had a cattle buying and pasture operation, and he also sent a lot of calves to Kansas. One of his Kansas buyers asked him to send some alligators on a load which he did. He put a cardboard box of small alligators in the front cab with the trucker and said, "Give this to the feedlot manager." When the trucker arrived, the manager wasn't there. The office lady took the box and made the worst possible choice. She called the Fish and Wildlife Service after she opened the box and screamed. To make a long story short, Ronnie was cited for transporting a protected species across about six state lines, and I think his fine was around twenty-five hundred dollars. He told me he thought he might go to jail. He said, "I'll tell you this for sure; I've shipped my last load of alligators." Ronnie was the picture of health but died in his forties.

Nic and Chris Lewis with Caroline and Kenneth under the boar's head.

Caroline and I with a 400+ pound boar.

Cousin Sally and Caroline in California.

Caroline in the California Ranch hot tub.

California black-tail buck.

Ken, Caroline, Betty and Wayne Gibson by the Madonna Inn Men's Restroom.

Jody, Mark, Caroline, Kennth, Annie Powell and Zeke Bricker at National Final's Rodeo

Elephant Rock on Carrizo Ranch

Dave and Diane Michelgard and Chic with Kenneth and Caroline at California Ranch

Caroline with calf from Precious

Part of the Carrizo Ranch elk herd

Precious and calf

Diane and Caroline at California Ranch

Carrizo Ranch antelope

Caroline and Jody Leslie heading out

–K–

Bachelor Housekeeping

K. S. Eng

Zeke is a hell of a cowboy
He is polite, young and single
When he smiles and nods to the ladies
They shiver, shake and tingle

He doesn't play hard to get
But he prefers the single life
He likes a big ranch, and the space
He has his animals, no need for a wife

Occasionally he yearns for a mate
Neighbors will bet and odds are laid
That if he is hunting for a female
His house is a wreck and he
needs a maid

It isn't that he is dirty
He showers and changes every day
But wherever he steps out of clothes
You'll know that's where they'll lay

He can be a pretty fair cook
But he never takes the time
He prefers Vienna sausage and chips
Plus Jalapeno dip, beer and wine

Dirty dishes are stacked in the kitchen
Wine bottles and beer cans everywhere
Tools and leatherwork cover the table
Guns and ropes lay in each chair

He is 30 days behind on his laundry
And it's stacked in one big mound
They found shorts, socks and Levis
And a missing Basset Hound

You've never seen so many socks
And yet not a pair that is clean
It's easier to buy them at Wal-Mart
Than throw them in the washing
machine

The females come in all shapes
and sizes
Many bring along their pets
If they tire of loving, scrubbing
and cleaning
They leave the ranch, but with
few regrets

First he takes them by the hand
And offer them chips and a drink
Then points them to the laundry room
And the dirty dishes in the sink

This year he's gained and lost
four women
And they've left a horse, a goat
and a pup
Zeke said, "You know it's not easy
To keep this house cleaned up"

The Cowboy Way in Frisco

K. S. Eng

We gathered our steers in California
Loaded & shipped them to Idaho State
With luck and rain the gains were good
And the market was looking great

We decided to stop in Frisco
And tour the city by the bay
Eat a steak & check out the girls
And show them the "cowboy way"

The restaurant I chose was a
little too fancy
Sal, our waiter was a little too cute
I said, "Bring a Coors & a rare steak"
Sal murmured, "Oh, you're
such a brute"

Sal said, "You act just like a beast
And you look divine in your jeans
Cancel your steak & come with me
Try my quiche & lovely greens"

As I ran for the door I heard Bill say
"Now it's my turn, follow me
I know a place we can get a steak
And genuine female company"

The waitress accidentally rubbed on Bill
She was "chesty", you know
what I mean
She said, "I'll show you some
new moves
I'm Lola, The Frisco Drag Queen"

When we finally got home to Idaho
We didn't lie, but we did deceive
When asked about nightlife & Frisco
We replied "incredible & hard
to believe".

-CHAPTER 24-

Additional Land Purchases in the '90s

An important land purchase we made was from Nic Lewis who was the token farmer near our California ranch. He owned about 2,500 acres adjacent to my ranch. Nic was born on the farm in the late 1920s, and had lived there ever since although he also had a home on the coast near Pismo Beach. Nic made his living dryland farming in that county all his life. To succeed, he had to be both smart and tough. At one time, he had a king-sized drinking problem which he later solved. Nic was smart. and if he liked you, he really liked you. And if he didn't like you, he didn't. He came to me twice to discuss selling parcels of the farm, and asked if I was interested. I appreciated it, and told him to have his lawyer do the paperwork. I bought over 2,000 acres. It was important to us because it was adjacent to the Carrizo ranch, and had a lot of elk and other wildlife on it. We discontinued the farming, but gave Nic an easement to continue living there as long as he wanted. This arrangement was fine with us because he was a good neighbor and good help. Sometimes when he was lonely, he and I and especially Caroline would talk for hours about anything and everything. Nic and his wife, Chris, stood up at our wedding at the ranch.

Nic was especially good working on windmills and old farm equipment. It made no difference how old it was, Nic could fix it. One day Loren and I came by his place and found Nic working on an old World War II army truck that hadn't been started for over fifteen years. We were going to the next ranch and I said, "Nic, how long do you think it will take to get it running?" He said, "It will be running by the time you get back," and an hour later, it was. He kept all kinds of old equipment around because he said you never knew when you might need spare parts. Nic died about a month before Caroline, and we loved him and missed him.

One of the strangest land purchases I made was adjacent to Fort Hancock, Texas which is an old Calvary outpost east of El Paso. It lays south of Interstate 10 and has a railroad track running through it. It is only about a mile from the Mexican border and a small border crossing point. A fellow who was a pretty good shade-tree carpenter was working for us at the New Mexico ranch. He mentioned that he owned 240 acres, and was going to lose it if he couldn't come up with some money. He asked if I might help him. I paid off his loan and charged him interest at two percent over prime with the option to keep a portion of the land. I had never seen the property, but thought its location was good because of the Interstate, the railroad and the proximity to a Mexican Border crossing point. Unfortunately, I underestimated his affinity for collecting used cars, equipment, junk, etc. and also his love for offbeat enterprises. The first time I visited the property, he was in El Paso. He said, "Just go to his house and a senorita will let you in. She doesn't speak English—watch out for the poultry." I knew he was in the racing pigeon business but I didn't know he also had fighting chickens. The lady let me in the house, and it was full of birds. I let myself back out and wandered around the property counting over forty used vehicles parked around. Unfortunately or perhaps fortunately, the house later burned down.

He found an old map indicating the land was zoned in half-acre lots and there were almost five hundred lots surveyed. He suggested we sell them to the local Hispanics who loved to buy a piece of land at an attractive price without much down and low monthly payments. He said we could sell them for $1,000 per half acre with $50 per month payments. We'd probably get the properties back within two years, and could resell them again. I told him the prospect of foreclosing on 500 Hispanics didn't appeal to me, and that when he paid me off, he could do what he wanted. Next, he started an Interstate 10 version of the Vietnam Wall on the edge of the property near the freeway. He said it was halfway between the East and West coast and advertised it in places like Soldi*er of Fortune*. For a fee, you could have the name of your friend or relative who died in the Vietnam War inscribed and put into the wall. The inscription was put on a cinder brick block, and I believe he had fifty or sixty sold when I saw it. His next venture was opening a flea market on our boundary near I-10 where he sold all sorts of antiques, nick knacks and junk.

With all the clutter plus the growing drug problem near the border, I told him I would discount the remainder of the principal, and give the land back to him. I did and hope he's doing well. Someday when I have some extra time, I plan to stop by and see how the place looks.

-CHAPTER 25-

1990s—Building a Yearling Business and Downsizing the Consulting Business

In 1990, Caroline was becoming more involved in my business and we decided to concentrate on running yearlings. We usually ran eight to ten thousand a year in Texas, Oklahoma and New Mexico. We brought small calves (three to four weights) and back grounded them at Mason County Feeders or other yards nearby. We grazed them on either wheat pasture or summer grass. Donny Stockbridge from Mason Cattle Company bought a lot of my calves, and while they weren't the prettiest, they had a good outcome. We fed the "tail end" we didn't sell as yearlings. Normally, we fed a couple of thousand head. Although our feedlot cattle were the "plainer end" they performed well. I delivered a lot of them on the fat futures until they got picky about how cattle looked. Their yield and grade was great, but they didn't fit the picture of what some thought a deliverable animal should look like.

Delivering cattle on the futures could be aggravating. We delivered one group to the packing plant at Dumas, Texas and they cut out an outstanding steer and called it a bull. It had a Mexican brand, so I knew it shouldn't be a bull and I told the grader he just had a big cod full of fat. The grader told me how ignorant I was, and I could see I might win the battle, but lose the war. I said, "Go ahead and kill him separately, but if you would, send me the testicles." There were none. The funny part of this story is they screwed up on the carcass weight and the animal yielded seventy-five percent and brought a big premium. Once again, it's not a perfect world, but it's good to be lucky.

Our main bank was Farm Credit West from Paso Robles, California and occasionally the bankers would come to check the pasture cattle. This was a bigger chore than they thought.

John Goldsmith would always tell Caroline, "If you wanted to screw us, you could figure a way so I don't know why I bother looking."

Caroline said, "John you're right but we'd never mistreat you," and that was the end of that story.

We sold a lot of our cattle on the Superior Video that was held every other Friday in Fort Worth in the old stockyards district. Caroline and I both loved the area. She loved it especially because there was a Leddy's western store next door. John Goldsmith and Bob Cox from California Farm Credit went with us to one of the videos where we had several lots of cattle for sale. We stayed and ate at the Stockyards Hotel which is a real treat, if you've never done it. Also the White Elephant Saloon across the street can be pretty entertaining and just as western as it was in the old trail driver days. The next day at the video, it occurred to Goldsmith this was a good way to check our inventory figures. John said, "I don't need to go to the pastures, I'll just go to the video at Fort Worth to get your inventory." John liked eating and sleeping at the Stockyards Hotel and partying at the White Elephant better than bumping across country in a pickup.

Caroline and I both loved going to the Superior Auctions in Fort Worth and we often spent two or three days there. We sold a lot of yearlings and calves on their video and they treated us well. Also it was a chance to meet old friends and to eat at great restaurants and bars. For Caroline, it was a western wear shopper's paradise. Leddy's Western store was next door, and was Caroline's favorite, but she also found other stores to spend our money. One summer evening we got to the Stockyards Hotel after the stores closed, but a pair of boots in the window at Leddy's caught Caroline's eye. I said. "If we have a good sale tomorrow, and they're your size, buy them." She came into the auction just in time to watch our cattle sell. She told me she bought the boots, and proceeded to give me a blow-by-blow account of the purchase. First of all, they were a pair of custom made boots, but the customer or ordered them never showed up when they were finished. Second, they were exactly her size, which was a sure sign that God wanted her to have them. Finally, she was able to get $150 off of the purchase price so it was a "win-win" situation. However, the purchase price was still $850. I told her, "God must have also known our cattle would sell well." I've still got the rose-colored boots, and I guess I'm looking for a Cinderella they might fit. In reality, I suspect I'll take the boots to my grave because I doubt I'll find the foot size or the personality to match the boots.

Caroline also loved to buy purses, and would buy a purse at each new location where we stayed. I put all the purses together after she died, and I believe there were 125. It's easy to find friends that fit a purse, so I've only got about 25 left. One of them is a Versace pink purse which she bought on our last trip to New York. She called to say she was looking at an expensive, but "one of a kind" pink purse. It looked like her and she loved it! So I said, "Buy it." Turns out it cost $2,000. It was worth every penny because she carried that purse every day the rest of her life through pastures, ballrooms and everywhere. She loved it, so it was money well spent.

Caroline was actually a very discriminate shopper and most items she purchased never went out of style. Abe Turgeon, his wife Ericka, Caroline and I were at a beach party in Hawaii one night and Ericka told Caroline she loved her new gown which was drop dead gorgeous. Caroline informed her, "Ericka, this is not a new gown." She then asked Ericka how old she was. Caroline said "Ericka, this gown is five years older than you are."

Buzz Williams, a Superior representative from Southern Colorado repped a lot of our cattle regardless of the location. He's a good rep who did his best to take care of both the buyers and sellers. He and I would usually sort the cattle together, and we had a few disagreements, but Caroline would help settle them. My favorite Buzz Williams story took place when we had several loads of cattle to be delivered on a Monday from a rough ranch near Tucumcari, New Mexico. The owner of the ranch called Buzz Sunday night and said he had the cattle gathered which was no small job, and he'd be ready at daybreak. Buzz had just had back surgery and was high on drugs. He told the rancher, the delivery was not until next week and turn the cattle back out. What a wreck. By the next morning when Buzz figured out what he had done, Caroline and I and the rancher were ready to put him back in the hospital. We gathered the cattle again and rescheduled delivery. From that time on I figured Buzz owed me one. Buzz, Caroline and I traveled a lot of pickup miles together to various ranches, videoing and shipping cattle. Buzz was a lot of fun and always said Kenneth, you don't know how lucky you are to have Caroline as a wife and partner. Caroline and I would often argue about how to sort and market the cattle, but after a few drinks we would go to bed happy. Buzz was wrong, I did know how lucky I was.

Tommy Gibson from Eastern Cattle Company bought most our cattle either direct or on the video. I know he ran into a lot of trouble later and I'm truly sad for all concerned. Fortunately, our experience with Tommy was always the best and that's how I'll remember him. As an illustration, one year he bought several thousand cattle from us on summer pasture in Northeastern New Mexico. We had a late winter snowstorm, and when it came time to gather the cattle, we were several hundred short.

I told Tommy, "I think the cattle are there, we just can't find them."

He said, "Gather whatever you can, and I'll wait for the rest."

We were running most of these cattle with Benny Gilbert. He called me that Easter Sunday and said, "Doc, it's not a pretty picture. We're in a hell of a blizzard and we've got cattle drifting all over northern New Mexico and Southern Colorado." He further said, "We'll do our best to gather them this fall."

I knew he meant it. In the meantime, he asked if I would partner with him to lease a ranch by Canyon, Colorado. Benny basically did everything himself.

"Benny," I told him, "you're going to kill yourself working this hard and driving so many miles. Let's leave that ranch alone."

Unfortunately, he leased the ranch, and in the fall after we had started shipping our cattle, he drove up there on Saturday night and back Sunday morning. On the drive back he missed a mountain curve and "bought the farm." Caroline and I both loved Benny, and Caroline did most of his bookkeeping. Benny was a hell of a cowboy who some said had a drinking problem so he stayed at the ranch. I never saw him take a drink and he really worked 24/7. We would go once a month to check our cattle and Caroline would check his figures and pay him at the end of each month.

In addition to running yearling operations, Benny had a great string of Hancock strawberry roan quarter horses that were flat out beautiful. Caroline and I had a couple picked out to take home that fall, but after Benny was killed that went by the wayside. Another fascinating story about Benny's horse business is he had problems with the bears coming down from the mountains and running his horses through the fences near his headquarters. Somebody told him if he got hogs, they would keep the bears away. He bought a group of Hampshire hogs and let them run wild around the area. He said, "I don't know if it was because the horses got used to the hogs or the bears didn't like the hogs," but he had no longer had bear problems. However, the hog population got totally out of hand, and because it was a closed herd they continued to reproduce straight Hampshire's except for one difference. Natural selection was taking place and each generation would have longer snouts. Picture it, perfectly colored Hampshire's with snouts six to ten inches long.

The end of the story was Benny got tired of the pigs and decided to take them to Guymon, Oklahoma to the hog slaughter plant. I don't remember the number, but it was enough to buy a new Ford pickup. I would bet if you check the back country forty to sixty miles north of Clayton, you might still find a few Hampshire hogs rooting around. When Benny was killed, Caroline and I were both devastated. To make matter worse, his family had no idea he was running so many cattle for us. Caroline said, "Why do you think I'm writing him the checks every month?"

After we had gathered the cattle nearby we were about 300 short. We hired a couple of cowboys with dogs and they took off for the mountains. They found three or four hundred more cattle that filled out our count. However, some of them were actually cattle that had been lost a year or two before. Age was not a big deal so it all ended well.

You've heard of outfits that are snake bit, and the Colorado ranch that Benny leased was one of those places. I think it belonged to a former major league relief pitcher who was losing it. Whoever took it over was short on help and Vic Ogle, Benny Gilbert's main man went up to gather the last of the cattle. There was a car accident near the ranch where an elderly couple had hit the cattle trailer and were killed. Shortly thereafter, the fellow managing the ranch was kicked in the head and killed while shoeing horses. Obviously, the place was bad news, and I was glad I declined to become involved. Nonetheless, Benny Gilbert had a nice family, and will remain one of my all-time favorite people.

Most of the country cattle we sold to Tommy Gibson were received by a rep from Amarillo, Texas. He was a good guy and I liked him, but if you gave him four head of cattle he would sort them four ways. We had a few disagreements but they worked out. The big problem was that he often wouldn't know where the cattle were to be shipped until we got them sorted and ready to load. Once near Wagon Mound, New Mexico, I had sold Tommy four loads of cattle, and we sorted them four different ways. I told him that was alright except we were going to weigh them up first for the pay weight. They were a pretty even group of cattle, but in spite of that, the rep insisted on lots of sorting. We finished and were waiting for the trucks, but they never arrived so we went to lunch. Then Tommy's man reported that he only had a place for two loads today, and he'll take the next two loads tomorrow. I said, "That's fine, but write me a check for all of them and take whichever loads you want." We went back to the pens and he couldn't remember which loads were which so all the sorting was for nothing. I got really pissed off, and Caroline warned me, "Don't mistreat him, we'll need him again someday." Once again, she was right.

A year later, Tommy's same man was receiving about two thousand cattle from us off wheat near Vernon, Texas. The morning started out well on the first couple of loads. I decided they didn't need me and went to a feedlot near Mason. Everything went to hell shortly thereafter.

Caroline called me later and said, "This is the sorriest bunch of cowboys I've ever seen. We're taking all the cattle into a nearby feedyard for sorting."

I said, "Whatever—but make sure we adjust the pay weight, and don't let the rep mistreat you."

She told me, "I'm not worried about him mistreating me. But it's so damn hot. It's 110°, and he's got a bad heart. I'm afraid he's going to die. You know we've got these cattle sold $10/cwt over the market so we need to go easy on him."

"Do whatever it takes," I said, "even if you have to give him mouth-to-mouth resuscitation or use a hot shot. Just make sure he gets the cattle delivered."

Caroline said, "I can do the hot shot deal but the mouth-to-mouth is out of the question."

Because we were running so many cattle in different areas of Texas and Oklahoma, sometimes Caroline and I would split up in the morning and ship cattle separately. I never had to worry about Caroline doing a good job because the truck drivers and cowboys were always on her side. One time we had a group of cows west of Wichita Falls, Texas. The person we leased the land from claimed to have about twice the acres as he had. I think he actually believed he was going to get the extra land leased, but he didn't. I told him to either get more land or we were moving the cows.

He said, "Come get them."

"We will be there tomorrow," I told him.

Caroline called Larry Ogle's wife who ran their trucking company and said, "I need eight trucks tomorrow morning."

"They'll be there," she answered.

We were staying at the Holiday Inn in Wichita Falls, and I was shipping wheat pasture cattle the next morning at Burkburnett, Texas. I told Caroline to go over and get those cows shipped, but stay away from the guy we leased from. He's got a reputation of being moody and unpredictable when push comes to shove.

I told her, "Keep a cowboy or truck driver close by."

We met later that night at the Holiday Inn bar, and I asked her if she had any trouble.

She replied "No. In fact the fellow turned out to be real nice. He and I talked for several hours, but I kept your pistol on the car seat where he could see it."

I said, "Caroline, that's a 9mm automatic, and you don't know how to handle it."

She responded, "I need to talk to you about that. Would you go out and check it, it's still on the front seat."

I went to the car and there was a 9mm Smith & Wesson automatic on full cock with a shell in the chamber. I came back and read her the riot act.

She explained, "That's why I wanted you to check it out."

What else can you say about a blonde? Bottom line, we got the cows gathered and moved without bloodshed. Caroline was not only smart and tough, she was absolutely fearless. This made everyone from cowboys and truck drivers to managers and CEO's respect her.

The best thing about our cattle business in the '90s was that Caroline and I could do it together or separately if need be. Often we would get up early, head our separate ways and meet again that night. I never worried for a moment about Caroline handling her end. Occasionally I suspect she'd do a better job than me.

Caroline waiting for trucks. So stressed she sneaked a cigarette.

There was a fellow near Wichita Falls called "Wildhorse" who ran a lot of wheat pasture cattle for us. He did a pretty fair job, but he was well named and he hated paperwork. That was another place where Caroline took care of all the charges and billings. He'd come into the bar at the Wichita Falls Holiday Inn with a shirt pocket fill of receipts and ask Caroline to sort them out. The first two days we were shipping, he showed up about one-and-half hours late while we had trucks waiting. Everyone knew the shit was going to hit the fan, and at the end of the second day, I was prepared for a "come to Jesus talk" with Wildhorse. However, Caroline got a hold of him first and led him behind a horse trailer and put the fear of God in him. The last day the cattle were gathered on time. Had I tried that, Wildhorse would have probably buried me.

Insofar as scientific advancements go, the '90s were a pretty slow time. We did get the first estrogen-Trenbolone combination implant cleared in the early '90s and also another ionophore (catalyst). Catalyst was probably a better ionophore than given credit, but its impact on consumption (or lack thereof) made it more difficult for some nutritionists to evaluate. I've always thought this was an underused product.

Feed prices spiked higher again in the mid-'90s, and once again cattle lost a lot of money. Overall, the decades of the 1980s and '90s were about breakeven for the feedlot cattle and fairly profitable for stocker operations.

In some areas there was a shift in the profit center in the feedlot business. The feedlot rather than the cattle became the profit center. This had a real impact on some of the larger feedlot operations and made custom-feeding less attractive to investors.

The other big change in the late-'90s and 2000 was the widespread but ill-fated feedlot expansion in the face of a declining herd. The result was a lot of excess feedlot capacity that is still haunting us. It has been estimated that we have twenty-five to thirty-five percent excess feedlot capacity. and we continue to pay for it in many ways. This is not a popular subject, but it an unfortunate example of ego getting the best of our judgment. The bottom line is that bigger isn't necessarily better.

Part was the result of new players in the industry. The big new gorilla in the room was Koch Industries that bought and expanded numerous feedyards in the Texas Panhandle. They also grazed several thousand wheat pasture cattle and were active in some of the same areas where Caroline and I operated. They had big plans, big money and big egos but that was not enough to guarantee success. After approximately four years, they quickly exited the industry and it's rumored their losses total three-quarters of a billion dollars. They were not the first energy company to try the cattle business, and in general, energy companies have not done well in the cattle industry.

As we got busier with our own cattle in the '90s, I continued to downsize my consulting pace. Sometimes this was a painful choice, and not popular with Caroline. When I gave up a number of my Northwest clients including Agri Beef and the Rebholtz family, Caroline was really aggravated with me. The Rebholtz patriarch, Bob, had a rare form of cancer, and he died at a young age. Their operation included three feedlots. Frankly, it was more than I could handle with our other personal projects. I told Caroline I needed to give this account up and turn it over to Bill Dicke. She was annoyed with both Bill and I. The Rebholtz family are not only successful people, but good people and good friends. Caroline was especially fond of their youngest boy. Paul spent a week with us at our New Mexico ranch. His youthful enthusiasm was a joy to behold. Unfortunately, he was killed in a plane crash shortly thereafter, which left Robert, Jr. in charge of the operation. They're still good friends with very successful operations, and I'm proud of them.

As we approached 2000, I told Caroline we were going to shift our emphasis to cow-calf operations, and I wanted to further downsize the consulting business. She went along with the idea, and the first decade of the new century treated us well. I've found it somewhat amusing that many cattle people continued to refer to me as a feedlot consultant. Except for special clients, I had substantially decreased my consulting activities in the 1990s and early 2000. This was not due to my lack of interest in consulting, but rather because of my other priorities—including cows and land. In spite of my downsized consulting activity, I still love the business and the communication with my peers. A day seldom goes by when I don't have several calls from consultants, feedyards, associated businesses, and so forth. I would be bored without them. Like the comparison of ham and the egg, it's great to be involved, but not committed. The end of the century, 1999, ended on a profitable note for all segments of the cattle industry even though the '80s and '90s were on average essentially a breakeven for the feedlot industry. Stocker operators faired a little better, and cow-calf profits were up and down. At the end of the '90s many in the industry were still talking about the ten-year cow cycle, but to the close observer, it was obvious times were changing.

Doing It Closer to Home

K. S. Eng

Sonny was a man of passion
He'd been that way all his life
That was his style and his fashion
He wore out several girls and a wife

He loved those big country ranches
Didn't want anything small like a farm
He always loved taking chances
He was reckless, that was his charm

When a market wreck wiped him out
You didn't see him cry or moan
He said "I still know what I'm about
I'll just do things closer to home"

Once in a fit of passion
He married a trophy wife
She liked the city and high fashion
Soon he longed for the country life

He pursued worldly pleasure and joys
And he coveted lots of space
Then he decided he had enough toys
And yearned for a smaller place

Bright lights and gold lost their shine
And he begin to slow his pace
Peace and love was his new
bottom line
As he prepared for the final race

He was relaxed when I passed through
I said "I hear you no longer roam"
He said "I still do just what I want to
I'm just doing it closer to home"

Lucky In Love

K. S. Eng

I've been lucky to love my work
And the friends that I have made
I would have done it all for free
Or if asked, I would have paid

I loved all the travel
No matter what the cost
I'd go to the middle of nowhere
And love it when I was lost

I've always loved the animals
Both the wild and the tame
I've loved the hunter and hunted
I loved them all the same

I have a special love for cattle
They have so much to give
They helped me make money
To live the life I love to live

I've always liked the ladies
A few of them stole my heart
But we usually ended up friends
Long after we drifted apart

But most of all I love you
You must have been sent from above
You're my partner, friend
and companion
You've made me truly lucky in love

-CHAPTER 26-

An Overview of Cattle Industry History

Perhaps this is a good place for an overall historical review of our industry. Compared to Europe, Asia and the rest of the world, the United States is a young nation with a short history. That's part of its romance and charm. Although there were significant agricultural events in the Eastern portion of the United States earlier, many say that the agricultural and cattle history in the Great Plains and Western U.S. began in 1862. This is 100 years prior to the beginning of my cattle business odyssey. The following is a brief review of the major historical events in the beef industry from 1862 to 1900.

1862-1900—Not Such Ancient History

1862—The Homestead Act opening up the Western U.S. to agricultural settlements.

1862—The Morrill Act—The beginning of Land Grant Institutions and experiment stations.

1866–1890—The Trail Drive era during which millions of wild and semi-wild cattle of Mexican origin were moved from Texas to Kansas, Missouri, Nebraska, Wyoming, Montana and other areas.

1867–1877—The beginning of the railroad era (including the completion of the transcontinental era of Promontory Point, Utah) that provided transportation for Texas trail herds moved from Kansas, Nebraska, Missouri to Eastern consumers.

1868—1878—The peak of the buffalo hunting for hide, meat and bones that ended with the near annihilation of the buffalo herd which once rivaled beef cattle numbers. This was an exciting but not necessarily proud phase of our agriculture development, and one of the early examples of abuse of natural resources.

1873—The collapse of U.S. banks plus too many cattle on the trail and at the railhead caused the cattle market to collapse. Cattle that had previously brought $35-$45 per head at the railhead declined in price to $5-$10 resulting in the first catastrophic losses of the trail drive era. An interesting footnote is that 100 years later (1973-74) what became known as "The Wreck" occurred, and again cattle losses were astronomical and cattle became difficult to market.

1874—Patents issued for barbed wire which eventually ended the open range era.

Late-1870s to early-1880s—The establishment of large ranches in Texas, New Mexico, etc. largely controlled by English and Scottish investors. In the early '80s the XIT ranch in the Texas Panhandle was composed of over three million acres, and by 1912, we had seen the last of the XIT cattle.

1886 & 1887—The great blizzards extending from the Texas Panhandle to Montana and the Canadian border that killed hundreds of thousands of cattle.

1876—Battle of the Little Bighorn which was the last major Indian victory.

1890—The Battle of Wounded Knee, the last major Indian defeat.

Twentieth Century History

1900–1999 (A Slow Start but a Fast Finish)

1900-1920

The beginning of the 20th century marked the end of the frontier area with Butch Cassidy and the Sundance Kid robbing their last train in 1901 and moving to South America. Some have said Butch and Sundance were the last westerners to deal successfully with the railroads. Also, train robber Blackjack Ketchum was hung in Clayton, New Mexico.

With widespread rains and good markets, agriculture was prosperous. Small fortunes were made in the Great Plains by breaking up grassland and planting wheat. The railroads expanded and promoted High Plains settlement and took credit for good rains due to the emission of steam from locomotives. They also suggested that plowing (turning the ground over) drew rain. Farm and ranch populations reached their peak of 31 million in 1919. Grain prices plummeted 50% in one year following the end of World War I. This was the beginning of rollercoaster prices which would plague agriculture for the remainder of the century.

1920s–1940s

While the general U.S. economy prospered during the '20s, the farm economy was in depression. The extreme drought and dust bowl in the '30s added to farmer and rancher misery. A part of President Roosevelt's "New Deal" was the first governmental attempt at supply management. This was the Agriculture Adjustment Act (AAA) that paid farmers, and killed their livestock if there was no market. Charlie Ball's firsthand account of "the day they shot our cattle" is a gut-wrenching firsthand description of government agents coming to the ranch, gathering their cattle and shooting them.

1940s and 1950s

Agriculture began to recover during World War II, and the government instituted rationing programs for food items such as meat and sugar. A black market developed for ration stamps, and in the West, rustlers frequently killed and dressed beef and horses at local slaughterhouses when beef supplies were short. In 1946, meat prices rose and price controls were renewed. The Brannon Plan designed by Agriculture Secretary Brannon began in 1950. It attempted to support grain and cotton prices at 90% of parity (1910–1914 prices levels).

In reality, few significant developments occurred in the beef industry the first half of the 20th century. However, beginning in the mid-1950s, this changed dramatically, and for the remainder of the 20th century we saw unprecedented new innovations in beef cattle nutrition and production.

Professor John Riggs reviewed "Fifty Years of Progress in Beef Cattle Nutrition" for the *Journal of Animal Science* in 1958. Always a gentleman, John was not prone to overstate or glamorize events. He concluded that "major contributions to the field of beef cattle nutrition in the past 50 years have been very few," and "Anyone who analyzes critically the rations used is bound to be impressed that rations fed today do not differ greatly in content from those used 25 years ago." The lack of progress is not surprising since we were preoccupied with the Great Depression and the Dust Bowl in the '30s and with World War II in the '40s." He believed the only significant event in beef nutrition was the discovery of Vitamin A and Carotene in the 1930s.

Prior to that, problems referred to as night blindness were common because many early feedyards used cottonseed products which were deficient in Vitamin A. He related an interesting story to me about judging when cattle had reached their "end point" in the feedyard. They would drive them down a lane at night.

The cattle that bumped into the fences because they couldn't see needed to be sold.

The growth promotant and implant era were ushered into the cattle nutrition scene in 1954 along with diethylstilbestrol—the first product approved. In 1956, '57 and '58, hormone implants were approved.

1960–1980

In the early 1960s the independent feedlot consultant business began to boom as did commercial feeding in California, Arizona and Texas. Live cattle futures began trading in 1964. The first-day volume was 191 contracts, and the close was $22.67 /cwt. The government also began providing food stamps to needy Americans.

Large amounts of grain in storage and the confusion related to government Ag policies resulted in chaos in the marketplace. The bans on trading with Communist countries were lifted in 1972. Grain companies purchased millions of tons of corn, soybeans, wheat, etc. They shipped these products overseas and obtained generous export subsidies as large as $60 per ton. Exports equaled one-fourth of the U.S. crop. Grain prices ranged from a low $1.40 for wheat and $1.14 for corn in 1971 and 1972 to over $4 a bushel for corn and $5.50 for wheat. The Brentwood Accord was signed and the U.S. went off the gold standard in early 1973.

In 1973 and '74, the economic event of the century for the cattle industry occurred and is remembered by many simply as "the wreck." The total cattle and feedlot numbers had increased to record highs in the early '70s partially as a result of tax shelter programs which enticed outside investors into the cattle business. The first hint of trouble was consumer boycotts because of high meat prices in '73s. As a result Nixon froze beef prices at the end of June. When the freeze was lifted on September 9, 1973, cattlemen assumed prices would rise dramatically. They had withheld cattle from the market to "fill the hole." Instead, cattle prices plummeted and continued to do so for the remainder of '73 and '74. Losses of $100-$200 per head were common, and because grain prices increased, some losses reached $500 per head. There was a tendency for feeders to withhold cattle because of the difficulty marketing them, and because the banker might not realize they were broke until the last cattle were shipped. A friend told me that if his last pen of cattle brought $10,000 per head, he might survive. Several experts predicted the end of cattle feeding with speeches were given and bulletins written predicting the end of grain feeding and the resurgence of grass-fed beef. I was on a dog and pony show speaking circuit at that time, and you would be surprised at the well-known individuals who predicted the end of grain feeding and feedlot industry. For feedlots, the silver lining in the '74 wreck was that feeder cattle prices plummeted. If you could refinance, the cattle that sold in mid-1975 when the market rebounded made lots of money.

1980s and 1990s

Government farm programs took a new direction and the Conservation Reserve Program (CRP) was initiated. It provided payments to take highly erodible land out of production for a minimum of ten years. Railroad mergers were allowed resulting in the closing of many country elevators which made it almost impossible to move grain from some Midwest areas to the West Coast.

The major beef cattle event of the '80s was the result of another misguided government program. In 1986, the dairy buyout went into effect, and resulted in a sudden and severe break in beef cattle prices. Those who survived the dramatic event and were able to reinvest in cheap feeder cattle recouped their losses in 1987. However, some never recovered and relations between the beef and dairy industry took a big hit.

Many economists and producers believed mergers and concentration in the grain and meat industry were depressing producer prices because fewer than ten major meat and grain corporations controlled the majority of the trade. Increasingly, the public viewed large corporate agriculture as arrogant and greedy. As a result, we were in danger of losing one of our greatest assets—the goodwill of the public. Ag Universities and Animal Science departments were not immune. Cheeke wrote a thought-provoking article addressing the shrinking membership in the American Society of Animal Science and related issues. He pointed out that although university research had improved efficiency of large corporate poultry production, the public did not view this as positive because of the resulting air and water pollution, as well as animal rights issues. Cheeke cited developments which had made animal and poultry science departments and sciences redundant. For example, while poultry production has increased dramatically since 1960, the number of poultry science departments had declined from forty-five to fifteen.

In the 1990s, numerous TBA and TBA-estradiol combination implants were improved. In 1996 there was an unexpected increase in grain prices resulting in severe losses in the cattle industry. Grain prices reached all-time highs with corn exceeding $200 per ton. Because there were ample outside cattle and feeder cattle, calf prices declined $30-$50 per cwt in a matter of months. Following the 1996 peak, grain prices declined rapidly and three years later they were down one-third from the high. In the late-1990s and early-2000s, feedlots began a rapid expansion program—especially in the Southern Plains—that proved to be one of the great tactical errors of our time. The attitude seemed to be that if we would build it, they would come. However, the custom-fed cattle investor was beginning to disappear, and the U.S. cowherd declined the next ten to fifteen years to lows not seen since the 1950s. The result has been a 25-30% excess feedlot capacity that has plagued us for a decade.

The following is a chronological list of major industry events and drug and implant clearances which occurred in the 20th century. It illustrates that although we got off to a slow start early in the century we really got rolling late in the 1950s and for the remainder of the century.

BEEF INDUSTRY MAJOR 20TH CENTURY EVENTS

1906—Federal Meat Inspection act followed Upton Sinclair's book "The Jungle" which exposed deplorable conditions in meat packinghouses, leading to formation of FDA.

1920—Packers and Stockyards Act led to formation of Packers and Stockyards Administration to regulate livestock trading.

1921—National Livestock and Meat Board (NLSMB) was founded.

1927—USDA began official grading of beef carcasses.

1929—Charles Goodnight pioneer cattleman (Goodnight-Loving Trail) dies at his ranch in Texas Panhandle.

1930–1935—Discovery of role of Vitamin A and Carotene in diseases referred to as cottonseed poisoning, fat sickness, night blindness, etc.

1933—Warner Bratzler Shear developed to measure meat tenderness.

1934— Passage of Taylor Grazing Act. Permitted organized grazing of federal lands for livestock production.

1935—Artificial insemination was first commercialized.

1940s—Congress begins funding beef breeding research and first heritability estimates are published in 1946.

1945—Use of urea and NPN as protein substitute began.

1950s—Dwarfism plagued the beef cattle industry.

1951—First successful transfer of a bovine embryo was made.

1952—First successful conception using frozen semen was accomplished.

1954—Diethylstilbestrol (DES) was approved for growth promotion in beef cattle. History of hormone product development as follows:

1954—Oral DES approved for cattle.

1956—Estradiol benzoate/progesterone implants approved for steers.

1957—DES implants approved for cattle.

1958—Estradiol benzoate/testosterone propionate implants approved for heifers.

1955—Ruy Kroc opens first McDonald's restaurant.

1956—Grain Sorghum hybrids developed. Subsequent sorghum production fueled feedlot boom in Texas.

1959—Delaney Clause established "zero risk" tolerance for carcinogens.

1960s—Rapid expansion of commercial feedlots in High Plains.

1960s—Use of independent consultants by feedlots increases. Started in late 50s in California by Dr. Brocklesby who was followed by Alego, Nofziger, Autry, Erwin, Cooley, Elam & Eng. Feed companies also expanded nutrition service to feedlot customers.

1960s—Feedlot industry develops or adopts steam flaking, high concentrate diets and feed fat.

1965—Yield grades were first used for beef carcasses.

1966—As a result of federal and state programs to irradiate the male screwworm fly, the United States was declared screwworm free.

1966—First Continental breeds imported to Canada from Europe.

1967—The box beef concept for further processing beef carcasses was introduced by IBP.

1960s—California net energy system developed by Lofgreen and Garrett. Became basis for computer ration formulations and performance projections.

1960s—Low level tetracyclines and bacitracins were approved for growth promotion and liver abscess control.

1968—Oral MGA approved for heifers.

1969—Zeranol implants (36mg) approved for cattle.

1972—FDA bans DES feeding.

1973, 75—Government freeze in retail beef prices followed by beef cattle market collapses. "The Wreck".

1975—First ionophore (Monensin) introduced.

1979—FDA bans all use of DES.

1982—Bovatec (ionophore) approved.

1983—The first beef embryos were split.

1984—Prepaid feed and tax shelters were question and many programs disallowed.

1980?—Tylan approved.

1980?—Virginiamycin approved.

1986—Beef check-off approved.

1986—Select grade approved.

1987—First EPD's were developed.

1988—Trade agreement with Japan resulting in record exports in 1989 and fat cattle market over $80/cwt for the first time.

1990?—Catalyst (ionophore) approved.

1990—Protein quality and requirements became a major research area and resulted in the introduction of several new concepts including protein solubility, rumen bypass, indigestible intake protein, digestible intake protein, etc.

1990—BSE problems in Europe cause concern in U.S. in 1997 and a ban is issued on feeding ruminant by-products to ruminant animals.

1993—E. Coli 0157:H7 outbreak in Northwest results in increased work on beef safety.

1990s—Numerous new TBA an TBA implants approved.

20th-Century Summary

In retrospect, we are an industry that celebrates its rich history and traditions. This is a small wonder since historical events range from exciting and exhilarating to depressing and terrifying. To the extent we learn from past events and mistakes, this is good. However, we are often accused of living in the past. The cattle industry is sometimes slow to admit mistakes and to accept changes in technology. Bill Pope relates a story of the old-time rancher who said, "I've seen a thousand changes in my lifetime, and been against every damn one of them."

Nonetheless, although the 20th century started slow, there was an unprecedented number of new developments during the second-half of the century that permanently changed and improved our industry.

As we progress into the 21st century, the development of new technology appears to have slowed, however new records have been set relative to prices, losses, profits, etc. These will be covered in Chapter 30.

-CHAPTER 27-

2000–2010—A New Millennium and a New Direction

Many expected widespread chaotic events to occur with the turn of the century to the year 2000 (Y2K). At our New Mexico ranch, numerous strangers from various areas drifted to the Gila Forest and Wilderness area prior to Y2K. I guess the fascination with the forest was the solitude and remoteness. Nonetheless, Y2K came and went quietly, and I suppose you could say things returned to normal except normal didn't last. In 2001, the September 11 attacks occurred and forever changed our lives and American culture. Changes in homeland security, airport security, government oversight of personal communications, etc. are now imposed on our lives, and while it may be necessary, personal freedom has taken a hit.

The best of times and the worst of times occurred in 2003. In October of 2003 fat cattle prices spiked to a high of $1.10-$1.20 per pound because of bovine spongiform encephalopathy (BSE) problems in other parts of world. Unfortunately, we then had a singular BSE cow occurrence in December 2003 that collapsed our market. In retrospect, I suppose the government and other support groups did the best they could, but the overreaction to the BSE problem was a shameful, embarrassing spectacle. The word BSE still sends tremors through the marketplace, as witnessed by the impact of the supposed BSE dairy cow in California in 2010. In reality, this was a non-event, but it still sent the futures prices down the limit a couple of days. Unfortunately, we are an industry that can be guilty of both inaction and overreaction and either can be devastating.

Waiting for Y2K

K. S. Eng

With the new year fast approaching
Some roll their eyes and say
It'll soon be the 21st Century
And they call it Y2K

When the clock strike's year 2000
All systems will crash they fear
But, I think there's a bright side
The wife's plastic may not clear

They say grain prices will tumble
Man what have they had to drink
Look at prices last harvest
How much lower can they sink

So what if meat prices plummet
They won't panic at the coffee shop
Cause when hogs hit $8.00 a hundred
How much further could they drop

If nutritionists formulas stop working
And the feedlot cattle grow thin
It can take them a year to fatten
And the market would be better by then

If vets battle an exotic disease
And no miracle cure is found
If they lose enough cattle
The market can turn around

I dreamed the clocks ran backward
And a 55 Chevy and Lover's lane
There I met an old girlfriend
But I couldn't remember her name

Suppose the earth would stop spinning
It might calm those cows that are rank
And we could finally get them gathered
Cause they're wanted down at the bank

Let the bank computers malfunction
And if they're unable to fax or phone
That means no past due reminders
And with luck, they'll lose our loan

If generators lose their power
And spark plugs lose their spark
No problem because since 1990
I've been operating in the dark

Perhaps my all-time favorite client story occurred in 2003 at Mark Adams, A2 Feedyard. Mark sold his feedyard in 2012 for a very attractive price. For the previous thirty-five years, he was a client of mine in Arizona and Texas. He is one of the smartest and best hands-on feedlot operators I have ever seen as well as one of the few that made and kept a lot of money feeding cattle. He has also endured some incredibly traumatic incidents the past few years including accidents that killed his wife-to-be, his stepdaughter, his best friend and feedlot manager. In addition, two of his other closest friends, Mike Wolfe and Larry Titsworth, moved to the big pastures in the sky. He survived it all, and is a lot tougher than I am.

Back to the feedlot story, after we shipped the last of our pasture cattle in 2003, Caroline and I had a couple of loads of plain but low breakeven yearlings left. Mark did a good job of feeding and selling plainer cattle, so I contacted him to say I needed to send a few loads to his feedlot.

He said, "I'm full, and can't take them."

I suggested, "I'll partner with you."

But he replied, "I'm still full."

To which I replied, "I'll even finance them."

Finally, he said, "Just send the damn cattle, and I'll find a place for them—and I don't want to partner."

The cattle sold in October 2003, and made $525 per head. This is an all-time personal record, and I quit feeding cattle for several years because I knew it would never be that good again. Mark and I still laugh about this. Although I'm glad he sold out while the timing was good, I miss him.

In the early 2000s, we had expanded our cow operations to California, New Mexico, Texas and Oklahoma. I was also in the market for additional land in Nebraska which Caroline wasn't thrilled about, but she finally gave in. She liked Nebraska, but hated snow. One night we were staying with Bill and Sharon Enrich in Norfolk, Nebraska and the forecast was for snow early the next morning. I had to catch an Omaha plane for an afternoon meeting in San Antonio.

Caroline kept saying, "Let's leave tonight and get a hotel room in Omaha."

I said, "No, Bill and I have wine to drink and guns to look at, and we'll leave early the tomorrow morning."

The next morning there were snow drifts everywhere but we left anyway. I had decided to go across the Missouri River at Decatur, Nebraska then down I-29 into Iowa. All we could see was blowing snow and a few snowplows—it was ugly.

Caroline said, "Kenneth, don't you dare have a wreck and kill us in Nebraska because I don't want to die hating you."

"Caroline, we are not in Nebraska," I informed her, "We're in Iowa."

"That's even worse!" she answered.

To make a long story short, we did make it to the airport, and the good news was there was no traffic. The bad news was you could barely see the road. Of all the stupid things I have done, that ranks close to the top.

I was so intent on building our cowherd because I sensed the industry was going the other direction with cow numbers declining but land prices going up rapidly. Looking at the table on page 275, you can see this dramatic change, and I call it the biggest transfer of wealth the agricultural industry has ever witnessed. The increased value in land and cows that occurred was truly breathtaking. It was obvious the old ten-year cow cycle was dead. That ten-year cycle was usually made up of a couple of years of profit, five or six years of breaking even, and two to three years of losing your butt. That's no way to sustain an industry. Either by design or because of better alternatives, it's obvious that ranchers have continued to liquidate cows. Some have converted pastureland to more profitable alternatives, and some took the money and ran. The relentless decrease in cow numbers and increase in land prices has been astounding the past ten years. Some suggest that, like the sheep industry, cow numbers will decline past the point of critical mass to the point where they are not a major factor in the U.S. Agri economy. I doubt it, but the love affair of cowmen for their cows regardless of the economics has disappeared.

The land cost per cow unit in many areas of the country has made profitable cow production impossible. In some areas, land prices range from twenty to thirty or thirty to forty and even fifty thousand dollars per cow unit. Even at today's low interest rates, this makes it impossible to pay for a ranch with cattle. If you believe the land will continue to appreciate because of recreation value, subdivision potential, or for other reasons, then it may still work. Some argue that this has always been true in the cattle business, and the only difference now is the number of zeroes involved. In the 1970s, experts believed the world's population would peak at six billion, and then gradually decline. However, we just passed seven billion, and many are predicting we'll reach nine billion people by 2040. Should this be correct, who knows what land may be worth in the future.

The year 2008 will go down as a wreck in almost all parts of the U.S. economy. Certainly, it was a bad year for the feedlot industry, but the stock market and U.S. economy in general fared even worse. The bankruptcy of Lehman Brothers, plus the collapse of the housing bubble were factors causing the economy and stock market to collapse. It was like what Joe Louis said about his opponents, "They can run, but they can't hide." There was no safe place to hide in that economy. I had a cattle business acquaintance who liquidated all of his cattle prior to the price collapse investing the money in the stock market. Selling the cattle was a good move, but when the smoke had cleared he had lost two-thirds of his stock market investment.

The first decade of the new millennium was exciting, but also occasionally terrifying. There was one area where activity declined substantially—this was the introduction of new drugs and feed additives. This decline may be related to a more stringent FDA, and the reluctance of drug companies to make new research investments in a volatile industry where numbers are declining. After years of research and FDA frustrations, Elanco was able to clear their beta-agonist, ratopamine, in 2003. The length of time required for FDA approval was probably a record. I was at an Elanco roundtable research review when they first floated the idea in the mid-1980s. At that time, conventional wisdom told us it would be cleared in the early '90s which obviously didn't happen. Intervet got their beta-agonist, Zilmax, approved the following year, and in 2007 Intervet also received a new clearance for a long-lasting combination implant called Revalor-XS which has gained substantial market share.

I believe the most important research breakthrough of the decade has been the advent of the residual feed intake (RFI) system for measuring feed efficiency. RFI is the difference between actual and expected feed intake of animals based on their body weight and growth rate. If you divide cattle into high, medium and low RFI groups, the low intake or the RFI group will often be more efficient because there is little difference in daily gain between groups. The ability to directly measure feed efficiency is a breakthrough because previously many assumed that high consumption and high gain animals were also the most efficient. The RFI system punches holes in that theory. There is much speculation concerning the mechanics behind low RFI efficient animals. Perhaps it relates to methane production, lower maintenance requirements, and so forth. Apparently RFI is also a heritable characteristic which means it may have a special role in our breeding programs.

A portion of the biofuels mandate of 2007 was a large increase in the cellulosic ethanol mandate of thirteen billion gallons to be reached around 2020. I doubt this mandate will be reached because it's difficult to break the lignin plant bond. Furthermore, cellulosic ethanol would not be as cheap as some think. Some have the misguided view that cellulose or roughage products are inexpensive, but on a nutrient density basis the opposite is true. Furthermore, the amount of roughage used to meet the mandate could result in a severe roughage shortage for the cattle industry.

The biofuels or ethanol mandate in 2007 resulted in a large increase in distillers' grains production. After the distillation process, approximately one-third of the corn used for ethanol production remains as distillers' grain. This byproduct is an excellent feed when properly used. Add to that the availability of corn bran and corn steep resulting from manufacture of fructose and starch from corn, and we had an abundance by-products available—especially in the Midwest. Because the dry matter feed value of these by-products is greater when they are fed on a wet or modified wet basis, feedlots near the source of production have a distinct freight advantage.

Some segments of the industry were slow to recognize the performance and health problems high sulfur in gluten and distillers' products may cause. Some mistakenly assumed that if polio or "brainer" problems did not occur, sulfur was not a problem. However, high sulfur can cause performance and subclinical health problems long before these symptoms occur. In many areas of the country sulfur levels have also increased in other ingredients. One of Newton's Laws of Physics is that you can neither create nor destroy matter. When removing sulfur from products such as coal and petroleum, it has to go someplace. I routinely run sulfur assays on several ingredients, and we're aware of some areas where the sulfur content of hay, for example, may be over 0.5%. Diets over 0.3 to 0.4 percent sulfur can impact health and performance if proper measures aren't taken.

Summarily, the major events of the decade were the relentless increase in agriculture land values and the decrease in beef cow numbers. Land prices have continued to accelerate and are at all time record highs. Beef cow numbers are at their lowest in over fifty years, and thus cow prices are at record highs. Ironically, many of the larger landowners and cow owners are one in the same, and the transfer of wealth to that segment of the industry has been "breathtaking."

I had decided to switch our cattle business from predominately a yearling grazing program to a cow-calf program. We already had cows on the California and New Mexico ranches, and I bought a ranch in Oklahoma, plus I leased several Oklahoma pasture properties. We built a cowherd of approximately two thousand head made up of a combination of black baldy heifers we had raised plus a lot of cows we brought out of drought areas in Arizona, New Mexico and Texas. My theory has always been that if the cows are priced right, buy them and find a place to run them later. Feed was relatively cheap in the early part of the decade, and we maintained many cows in dry lots for several months before moving them to pastures. Our timing was good, and the cows were profitable every year for the first decade. This is not to say we didn't have plenty of problems with drought, unexpected blizzards, theft, etc., but the market timing was good enough to overcome these events. Sometimes I think droughts and blizzards follow wherever I may go. We had several very dry years in New Mexico, and late in the decade we had two years of record blizzards in Oklahoma. I'd rather have neither, but if push came to shove I'd take a drought to a blizzard.

Because of the high price of land and droughts, we used dry lot cow production programs for several years. I kept quiet about it because we didn't want competition buying cows. Also, I figured many would assume I was crazy. Caroline preferred the cow business since her Dad had ranch, and also he always had white Brahmans. Caroline loved Brahmans—preferably the white ones—and regardless of their temperament we always kept a few. One year we were at the Templeton California sale, and I remembered it was her birthday. I had forgotten to buy a present. They ran a poor Brahman cow with a sorry little Brahma calf into the ring, and I bought them. As we were leaving I said, "Caroline, you're birthday present in penned behind the sale barn," and she couldn't have been happier. She named the cow "Precious," and that she was. She had ten straight calves before retiring. I kept Precious until she died January, 2014.

Precious and calves

In late 2009 we were trying to buy a ranch in Nebraska in addition to keeping track of all of our cows. I told Caroline I needed to let the Lasley account at Stratford, Texas go. Walter Sr. and Jr. Lasley had been clients for over twenty years and Caroline loved them and their feedlot people. She got really mad at me but I convinced her I didn't have time to do the job I should for them, and they would be better served by another consultant like Jim Simpson. The first time I took Caroline to the Lasley Feedlot, Walter Sr. was still alive.

When he was busy, he could be pretty abrupt. I warned Caroline he probably wouldn't have much time for small talk, and she said "I hear that he and his son are big Aggie boosters, so I'll bet we'll get along." It was amazing. She'd spend most of her time talking to either Walter Sr. or Jr. about Aggie football and politics. That was another advantage to our relationship because Caroline liked the people I did and they liked her.

That brings us to the end of the decade and perhaps this is a good place discuss some of the special employees I've had. Like every company, even a small one like ours, you will occasionally have an employee that doesn't work out and disappoints you. Fortunately, they were in the minority. We had a manager for a short time at the California ranch that was capable, but had a drinking problem. We weren't at the ranch much so he had limited supervision, and he started making a lot of 900 phone calls on our phone. When we got the monthly phone bill it showed about six hundred dollars in charges for 900 calls. We contacted the phone company and said there must be a mistake. We had already asked our foreman and he denied the calls. The phone company asked if we could identify his voice as they tape all 900 calls. They began playing them back and sure enough, it was our foreman. He was calling Madam Cleo for advice on career, love, and whatever else, and I guess she told him what he wanted to hear. The worst part about the calls is that most of them occurred in the middle of the afternoon, so he was not only running up the phone bill, he wasn't working. Anyway, for any of you tempted to make 900 calls be forewarned, they may be recorded.

I've had a lot of good office staff. Working in our office requires everything from typing and filing to bookkeeping, banking, etc. We like our people to be flexible, easy to get along with, work hard and hopefully smart. In turn, we pay well and have flexible hours. Best of all you won't have to put up with negative people, at least not more than once or twice. My theory is life is too short to drink cheap wine or associate with negative people you don't like.

When I first started my own business, my then wife Pat was a one-girl office, and she was good. She had all the necessary office skills, and was probably smarter than me. However, she kept going to college after her original degree in microbiology and ended up with a couple of master's degrees in education specializing in reading disabilities. She wanted to teach rather than be my office manager—I couldn't blame her. Her students were the real winners.

My next office manager was also named Pat. She was not only good help, but became a close personal friend. Pat had a great sense of humor, and she and her lawyer husband had moved to Casa Grande, Arizona from Wyoming. I believe he was the Wyoming Attorney General, and he took over the City Manager's job in Casa Grande. He left a year or two later for private practice. I've come to appreciate lawyers later in my life, but at that time, Sam was one of the few lawyers I liked. He was a no nonsense but fun fellow who had some interesting friends. Wherever I've had a office, it's always been built onto or part my house and that was the setup in Arizona. Sam had one particular friend—Billy—who would come by and visit Pat. He was an exact replica of a young Ernest Hemmingway, and had a charming-rogue personality. He was from Oklahoma, and didn't seem to have a job, but had plenty of money. He made a big impression on the local ladies. When he arrived in Arizona, he and Sam had a habit of disappearing for a week or two. Pat thought they would go to Mexico and do whatever guys do. She thought Billy was a gambler and lived by his wits.

Pat, who was never ill, started feeling bad. She thought she had the flu, and just toughed it out. When she finally went to the doctor, they found cancer, and she was dead three weeks later. Everyone who knew Pat was heartbroken, and Billy, who was Sam's best friend came for the funeral. I left a day later, but Billy stayed with Sam. About four days later he called our house at four in the morning. He got my wife on the phone and said, "Sam and I have had a big fight and I need a ride to the Phoenix airport right away, can Kenneth take me?" I wasn't there but Pat got our two girls out of bed, loaded them in the car, picked up Billy and dropped him at the Phoenix airport about six in the morning.

We were part owners in a travel agency at that time, and a few weeks later the lady managing the travel agency said the FBI are here, and they want all of Sam's airline travel records. I exclaimed, "What the hell is that all about?" I called Sam and he said it's a misunderstanding but let the FBI see them. To make a long story longer, it turned out Billy was on the FBI "Ten Most Wanted List" as a bank robber. Their *modus operandi* was to kidnap the bank president and his family, keep them overnight, then get into the bank early when the safe could be unlocked. They accused Sam of being part of the robbery ring believing that he set up their "safe houses." Sam denied it, but about a month later he killed himself with a .357 Magnum. I knew Sam pretty well, and I'll never believe he committed suicide. If he was involved in the bank robbery group, I believe they thought he was the weak link, and they gave him an easy way out. I'm told Billy was later captured and sent to prison.

Occasionally my daughters and I will tell their mother that when she took Billy to the Phoenix airport, he was probably on the way to rob a bank and she was therefore an accomplice. My daughters and I think this is pretty funny, but Pat fails to see the humor.

My next office manager was Jess Shearer, and she was a real gem. Her strong suit was as a bookkeeper because she was meticulous, tenacious and honest. She was a great help. I again had my office in the house which by then was a small condo in Casa Grande. We did a lot of dove hunting, quail hunting, etc., and I enjoy cooking so we had several neighborhood parties. One evening my Phoenix lady banker was coming down for dinner. My crockpot was broken and I called Pat, my ex-wife by this time, and asked if I could borrow hers. She said "Sure, I'll bring it over," and she wanted to know what and who I was cooking for. When Pat left Jess said, "Kenneth, you're an idiot," and it turns out she was correct. Pat just happened to stop by the condo later that night and it was embarrassing. The next morning Pat called to say, "I'm sorry. I was so stupid." I said, "I'm sorry, too." Then she asked if she could borrow my Blazer to move some books.

When she came by for the Blazer, I had several guns in the back from a dove hunt and we were carrying them back into the bedroom. I had a 12 gauge Browning automatic, and made sure it was unloaded, but apparently there was a shell stuck in the chamber. My procedure is to point the gun straight up in the air and pull the trigger. Instead of a click, there was a roar of a 12 gauge. I can tell you this, the sound of a 12 gauge in a small bedroom is deafening, and the big hole in the ceiling with insulation falling out on your head is embarrassing. Pat was petrified, I felt stupid, and we forgot about Jess working in the next room. The next thing I heard was Jess asking, "Is everybody okay or should I call the police?" Fortunately, that wasn't necessary but I told her I probably should be locked up for stupidity.

When I bought the New Mexico ranch, I vacated Arizona and built a nice office on what had been originally a Forest Ranger house. Caroline soon took over the duties there. She hated typing but was a great bookkeeper and banker. Furthermore, she had an opinion and wanted to be involved in every business aspect—which suited me just fine. The best thing she did was to hire Annie Powell who has worked for me for close to twenty years. Annie and Caroline got along great, and when we left, Annie stayed there with the New Mexico office. Caroline and I moved to San Antonio because it had good plane connections, it was close to many clients, and also relatively close to her family in the Bryan-College Station area. We had a San Antonio office, but travelled so much that Annie did most of the office work. She and Caroline would communicate constantly by phone, fax, computer, etc. By then we had a pretty complicated multi-state business in research, consulting, cattle, land, and so on. Those two kept on top of it. Caroline and Annie were the true definition of 24/7 people. Even now Annie often starts work at three or four in the morning.

Because of the increase in land transactions and the fact that the Biolite® mining business was going well, she added two more staff at the New Mexico office—Fran and Sharon. They stay in New Mexico, but Annie spends about half her time in San Antonio, Mississippi or travelling with me. Violette is another good friend who is in charge of my wardrobe, the toughest job of all. She and Annie are best friends.

Jessica Shearer at the Arizona office

Perhaps this is a good time to mention my mining venture, which is built around a very unique calcium-clinoptilolite zeolite. Its trademark name is Biolite® and we've been researching and marketing the material for approximately twenty years. I've conducted over thirty feeding and research trials on the product, and it is probably the best researched new feed ingredient on the market. The research has been conducted by both university and private groups, but we've discontinued publishing results. Whenever we made the research results public, I found there were copycat groups that would claim to have a similar material. Even though their claims may be incorrect and false, I was afraid this would have a negative impact on the general perception of Biolite®. Biolite® is an extremely good rumen buffer, binds several toxins (aflatoxin and ergovaline) and reduces ammonia emissions therefore increasing the nitrogen content of manure.

This book is not meant to be a Biolite® advertisement but it is now widely used in several states and by a number of feedyards and Nutrition consultants. Annie is in charge of the Biolite® marketing, and if you're interested, she can be reached 575-743-2100. One other side note, Biolite® is part of a Zeoponics media to grow plants in orbiting space stations, and it has also been used to bind certain radioactive waste. There is limited economic significance to us, but it's an interesting sideline.

Mark and Jodi Leslie managed the California ranch. They worked for us over ten years and were great employees. They treated the ranch like their own whether it's regular ranch work, elk hunting, or whatever, and we miss them.

Zeke Bricker worked for us several years in New Mexico, California and Oklahoma. He's a confirmed bachelor, so he didn't care where he hung is hat, as long as it wasn't close to town. Zeke is the son of Ed and Mary Janet Bricker, and the baby brother of Ed Bricker, Jr. Ed and Mary Janet were from Arizona, and Ed managed the Hughes and Ganz feedyard in Arizona and later managed outside cattle for AzTx in the Texas Panhandle area until he retired. Eddie now has a feedyard in Alfalfa County, Oklahoma, and just brought a ranch near Medicine Lodge, Kansas. Eddie Jr. and his wife, Manda, have two young daughters who are also going to make good hands. Eddie previously had a feedyard south of Chickasha, Oklahoma, and he and Zeke both worked for me at times. Zeke was a hell of a hand, but hard on horses, houses and himself. While working for us, he had several wrecks and broken bones. Eventually he was so crippled up at the age of thirty-five that he could hardly mount a horse. We retired him and since, he's had some back operations that have improved his posture and looks. He now looks about sixty rather than seventy—but in reality he's only forty. Eddie and Zeke were great hands, and if you needed to gather cattle from rough country or brush, they could do it.

Several years ago, I had about a hundred head of heifers on wheat pasture that I was going to breed. Eddie and Zeke went out one morning and found a large buffalo bull in the pasture. We have no idea where it came from, however, they roped it, drug it out and turned it loose. I forgot all about it until about nine months later. We'd pregnancy tested the heifers, and had sent them over to Jeff Dees who was southwest of Chickasha to pasture them and calve out. He called me a couple of times with bad news saying we'd had some late-term abortions in the heifers.

He then called me a couple of days later to say, "Things are looking better, we have two live calves." He also asked what kind of bulls I had used.

I said, "Just normal, small-frame Angus."

Jeff mentioned the calves looked a little different. When Caroline and I looked at the heifers a couple of months later, one of the first things I saw was two beefalo calves.

I said, "Jeff, have you ever seen a beefalo?"

"No," he answered.

I told him, "Yes, you have, you just don't recognize them. There are two right in front of you." Abortions are common in buffalo crosses.

One last comment on the first decade of the new millennium, while the cow-calf business and stocker business was consistently profitable, profits were infrequent for feedlots. Feed prices were erratic but trending up. Occasionally certain by-products were outstanding buys, and as a result, if you could think outside the box, there were some occasional decent feedlot profits.

Jessica and Ken at Casa Grande Airstrip

–K–

My Cows Aren't Mad But I'm Getting Aggravated

K. S. Eng

Ever wonder why cows are happy?
Well now it's plain to see
I know why they're content
Cows never watch TV

Cows never read a paper
Or watch the TV crews
They just lay around the pastures
While we worry about the news

The news of a "bleeding heart"
Or an activist who is sad
Because we modified a soybean
Or one cow that may be mad

They forget the world's hungry
Or starving children who may die
They question the farmer's motives
And "bad mouth" our food supply

They fret about animal rights
And now we hear their cries
About a kid who is too fat
Because of his French fries

They ignore the hungry and homeless
They'd rather pick on a farm
That produces food they can overeat
That's their idea of risk and harm

Even if odds are one in a billion
If it's food, no risk is too rare
They like to cover the unimportant
If they can find someone to scare

They don't report, they interpret news
They love to rave and rant
They love any headlines
They can give a negative slant

Food producers should take a vacation
And go to some retreat
Then news reporters could call
When they decide they want to eat

I'm swearing off TV and news
I think what I'll do for now
Is find a peaceful pasture
And hug a happy cow

We're Still a Pair (Happy Anniversary)

K. S. Eng

The boots and horns are old
The spring flowers are new
It's a picture that tells a story
That's a lot like me and you

Our boots are worn at winters end
And horns are shed by elk and deer
The grass greens and flowers bloom
And we're a pair for another year

The cows are dropping calves
The horses shed their hair
Birds return from the South
And you and I are still a pair

The birds are building nests
The coyotes are digging a den
Nature's chosen are choosing mates
And we're a pair again

Spring is my favorite season
There's a rhythm in the air
Plants and animals come alive
And best of all we're still a pair

Photo by Jodi Leslie

-CHAPTER 28-

2010–2012
Only Three Years But Seemed Like Forever

I hope you won't find this chapter too depressing—and it's not a plea for sympathy on my part. Rather, I want to cover details of my depression following Caroline's death in hopes it may help others deal with similar problems.

The beginning of 2010 started about the same as previous years. I was in the mood to buy a Nebraska Ranch, and Caroline, who hated cold weather was trying to discourage me. We looked at two ranches, one near Broken Bow and the other on the Niobrara. I was doing my best to sell her on the idea. We went to the National Cattlemen's Beef Association (NCBA) meetings, and the rest of our time was spent on the routine land, cattle and consulting business. Beginning in March, it became obvious that there was a subtle change in Caroline's energy level. She was normally a 24/7 type of person, and only occasionally would treat herself to long mornings in bed.

Beginning mid-March, she often stayed in bed until almost noon, and would watch TV, read or talk on the phone. She also had some unusual pains in her muscles and joints. I knew something was wrong, so I called our family doctor who immediately sent her to the cardiac ward. We spent a day there, and when we were through, they gave her a clean bill of health plus a bill for $38,000 that fortunately was covered by insurance. They concluded there were no problems, and she was in incredible shape for a woman approaching seventy. I could have told them that. We went to Dave

McClellan's annual client get-together in June which is always fun. Dave and Judy put on a great meeting each year and have great clients. We also went to Bryan-College Station, Texas several times to visit Caroline's family. We spent more time with her mother who was ninety-two years old and in a nursing home. Her mother's mind was not working well, but if you didn't know her, you couldn't tell. She moved almost as fast as Caroline, and in her mind, she believed she was managing the home. The people working there were understanding, and let her do whatever she pleased in the kitchen, laundry, recreation room, and so forth. However, they tried to keep her out of the office and the books.

I'm thankful that in the weeks preceding Caroline's death she and her mother had several good visits. This was unusual because Caroline and her mother often didn't get along. I think her mother resented the fact that Caroline was her Daddy's girl, and she spent most of her young years working cattle and horses with him. Fortunately I had more patience than usual, and told Caroline not to worry about work because it would get done. We relaxed more, talked a lot, and she said several things I had never heard before. Caroline did not like to talk about death, funerals, cemeteries, or the like, but she said several times she was glad she found a spot for me in their family plot at Shiloh cemetery. In fact, she was so thrilled I told her, "I hope I'm dead before you plant me." If you're a cemetery aficionado, the Shiloh cemetery is a beautiful place in the country in Robertson County. One of Caroline's original ancestors donated it to the community around 1850 and it has lots of interesting graves and tombstones. There are a large number of Confederate war veteran graves—I even found a tombstone for a War of 1812 veteran.

Caroline also wanted to talk about our funerals which was really out of character. She wanted the services in the Baptist Church in Bryan, but didn't know who should conduct them. I suggested Baxter Black, and Caroline liked that idea. Baxter did a wonderful job at her service as did her cousin who is a Baptist lay minister. As some of you may know, Baxter has become a pretty religious person, and I now call him the Reverend Dr. Black. His eulogy is at the back of this chapter.

About a week before Caroline's accident we were sitting in the living room one night sipping wine when she said, "We should repaint this room." Then she added, "I don't think I'll do it because if you die, I couldn't live in this house without you." I asked her what made her think that I would die first. She said, "I don't know, but whoever dies first, will be the lucky one." She sure got that right. A week later in late June she walked by my desk one evening and said, "I feel so tired and logy I think I'll get in the swimming pool." The next time I looked up, she was floating face down. I jumped in, pulled her out, gave her CPR and mouth-to-mouth and called EMS. They got there quickly, and I had gotten a slight pulse back. However, within a day it was obvious she had permanent brain damage.

I began to worry she might survive in an invalid state, and would never forgive me for saving her. The doctors asked if we had a Living Will, and I said, "Yes, but I don't know where it is." The girls went back to the office, and found it on top of her desk. Go figure! The last night we were at the hospital, the doctors asked if I wanted a minister since they kept one on call. I said that would be good, and they sent a fellow I believe was from the Caribbean Islands. He started by praying, then began chanting and dancing.

I said, "That's enough," thanked him and gave him $20.

Ron Crocker was with me, and he said, "You kind of cut him short, didn't you."

I said, "Yeah, I was afraid his next move was going to be pulling a chicken out of his trousers and wring its neck." I'm sure he meant well.

Looking back Caroline and I had met briefly, and casually around 1970. When we parted, she said, "Kenneth, someday I'm going to marry you." I knew it wouldn't be any time soon, and like John Wayne, I thought "that'll be the day." We were too similar—both reckless, a bit on the wild side, headstrong and impulsive. I figured together, we would have done a lot of damage to each other.

Out of the blue eighteen years later she called to ask what I was doing, and as the saying goes, "the rest is history." I was unattached and had mellowed, and she had too. We each loved travel, ranches, cattle and worked and played hard. Ours was not a "sugar and spice and everything nice" relationship. I suppose you could describe it as a "lusty all-in" relationship between two strong willed, somewhat selfish but generous personalities. It would not have been a relationship for most people as we probably averaged at least one good argument a day (usually about business). At times we would yell and shout, but we seldom went to bed mad. We each had a lot of flaws, and we weren't perfect, but we were a perfect pair. During our last twenty years, we were apart only two nights.

The official verdict for Caroline's cause of death was drowning because they found water in her lungs. However, the coroners are extra busy on weekends in San Antonio, and I suspect had they looked closer they might have found other problems. Caroline was a good swimmer and I personally suspect she suffered a stroke while she was in the pool. Dead is dead, so it really doesn't make any difference. I discovered that the sudden death of someone extremely close is far different than the death that has been delayed, overdue or a long-time coming. My father suffered from terminal cancer in all parts of his body his last several years, and my mother suffered from Alzheimer's her last four years. In both instances, I was relieved when they passed because their suffering was over, and they had to be going to a better place.

Since Caroline died, I've had many people call with good advice. Also people called telling me they are suffering from chronic depression, and asked what I would recommend. Some of the people calling really surprised me as I had no idea of the problems they were dealing with. The best piece of advice I got was two days after Caroline died when my neighbor Charlie called to ask if we could visit when I had time. I said, "No time like the present," and we talked.

He said "Kenneth, one of the things that's will help you get through this is you'll find people who want to relate a personal tragedy of their own. While you won't feel better because they've suffered a tragedy, you'll feel better because you'll know they survived."

I asked Charlie what his tragedy was and he related a story that chilled me to the bone. It involved the murder of his twenty-year-old son. He said for five years thereafter he considered killing someone or committing suicide every day. Charlie is like the Rock of Gibraltar, and his story and advice really helped. Less than an hour after Charlie left, Bob Josserand called to relay his family's sympathies.

I said. "Bob, you had a similar drowning incident with a son didn't you?" He went on to recount the drowning of their son at Lake Meredith where it took them four days to recover the body. He said each day they hoped and prayed they would find him clinging to a rock or a small island, but it wasn't meant to be. He said it was the worst four days of his life, but Bob and his family survived. That gave me a glimmer of hope.

For the next year, I didn't just think of Caroline every day or every hour of the day, I thought of her every waking minute of each day. Had it not been for frequent calls and visits from friends, I doubt I would have made it. I tell those who ask that what helped me survive was numerous friends, plenty of wine and work. Also, I found that I began to sleep more than I ever had. If not busy, it was not unusual for me to sleep twelve to fourteen hours a day which I suppose is a form of escape.

I'll Get the Gate

K. S. Eng

I dreamed I saw an empty saddle
Nearby stood a paint and a gray
They were waiting for their favorite lady
Who always brought their morning hay

The pick-up passenger seat was empty
Where you sat when we used to date
I joked you were there to ride shotgun
But it was really to open the gate

Next I looked in the office
Near the desk was your empty chair
On the desk were some unpaid bills
And a ribbon from your hair

I dreamed of our country cemetery
Just why no one knows
And there was a fresh dug grave
On top lay a single red rose

Slowly I reached for your side of the bed
Thank God your warm body was there
But I couldn't go back to sleep
I kept re-thinking my worst nightmare

That morning we were in the pick-up
We got started a little late
I said "Honey just sit and look pretty
From now on, I'll get the gate"

When Baxter, Annie, my daughters and I were going to the cemetery after the funeral Baxter said, "Let me warn you Kenneth, you're not going to sleep well for a long time."

I said, "Baxter, you may be right, but what makes you so sure."

He said after his former wife divorced him he didn't sleep for four years.

I'm improving, but it's strange the things that still bother me and bring back memories. I had to sell a couple of ranches where we were especially close. Also, there are certain places I can't go without her, including Vegas and New York City because Caroline was built for the bright lights and Big Apple. However, I've stayed busy—mostly trading land and cattle—since she died, and fortunately this has gone well. Unfortunately, I sometimes feel guilty about it because she's not here to share my good luck.

Perhaps what helped the most is that I started the Dr. Kenneth and Caroline McDonald Eng Foundation, mostly to fund beef cattle research at Texas A&M, Oklahoma State and the University of Nebraska. Also, I've help fund building projects at Wayne State College, and given out several scholarships in my hometown area. We also fund the PNC poster session.

Money as such has never meant much to me, and it seems I have more than I need. I've donated two million dollars to the foundation, and may add more later. Giving back can be therapeutic. Much of the research funding is on improved beef cow efficiency, and we began holding annual symposiums on that topic beginning in September 2013. We plan to rotate the meetings. The first one was held at University of Nebraska. The second will be cosponsored by Texas A&M in San Antonio, Texas, September 18th and 19th 2014. The University of Nebraska and Oklahoma State will also participate.

For whatever it's worth, my advice to those suffering from chronic depression is to spend lots of time with friends, work, drink wine if you're so inclined and sleep a lot. There's also a big place for religion, and I'm working on this. Psychiatrists and pills haven't helped me, but I suppose each person is different. One last piece of advice, don't try to forget your departed loved one. I frequently get calls from people who want to relate how they miss Caroline or Caroline and I together. Originally, it bothered me. I now find I'm much happier when I remember Caroline and the good times. I've read some "how to cope" books that say the least desirable and most painful relationship to deal with losing is one which is extremely close and both people are dependent on each other. Inevitably, the survivor will be extremely sad and depressed. This is possibly true, but I wouldn't trade our twenty-plus years together for all the gold in Fort Knox, and, I'd trade everything I own to have her back.

Now back to our industry basics. The years 2011 and 2012 were more of the same insofar as cow liquidation and land value were concerned, but there was one very important new dimension. We had:

1. Continued beef cow liquidation

2. Continued land value increases

3. A severe drought in the Southern Plains in 2011 and in the Central Plains in 2012.

October 2010 through October 2011 was the hottest and driest twelve consecutive months on record in several areas of Texas, Oklahoma and New Mexico. I put together another cowherd in April and May 2011 in the San Angelo to Sweetwater, Texas area. I went through the normal progression of those enduring a drought from, "It's got to rain soon," to "It's got to rain some day," and finally, "It isn't ever going to rain." Doctor Wes Bonner managed my Texas confinement cows. An unexpected problem was coyotes that were also hungry because of the drought. They killed an average of one coyote per day for sixty days.

I bought a thousand head of either heavy-bred cows or cows with small calves. Most of the cow-calf pairs weighed around 900-950 pounds, and had they been in good shape, they could have weighed 1300-1400 pounds. It never rained, and we ended up keeping the cows in confinement for approximately sixty days. When ration prices went up significantly, I put wheels under them, and moved them to Nebraska. I also moved most of my Oklahoma cows to Nebraska. We didn't have much extra grass up there, but the by-products were relatively cheap, so we could maintain them in dry-lot for approximately $1 per head per day less than in Texas. I also bought several Nebraska cows. As a result of a good calf market in the fall of 2011, and a good packer cow market in the winter of 2012, it worked out well. Unfortunately, the drought hit Kansas, Nebraska, and Wyoming and other areas in the summer and fall of 2012. Conditions obviously changed for the dry-lot ration costs in Nebraska. It proves the old adage that "if you keep trying different places at different times, you will always be in trouble somewhere."

It has been a continual learning process, but over the years semi-confinement cow programs have treated me well. One advantage of a confined cow operation is that you can blend the diet on a least cost basis, and program feed it to the required levels. This reduces feed requirements. Also calf weaning and preconditioning is a piece of cake because the calves will be eating with cows prior to weaning. Finally, it's much easier to apply all types of technology in confinement. whereas it's nearly impossible in places like my New Mexico forest lease with five cows per section.

In March of 2012, "Pink Slime" received widespread coverage in the media, especially on ABC News, and also so-called "reality shows." It blindsided us in the market place. "Pink Slime" is, in reality, lean finely textured beef (LFTB), a category of beef products using high tech equipment to separate lean meat from fat, because doing it by hand would be impossible. LFTB products prevent the waste of valuable nutritious and safe beef and were approved by the FSIS and USDA for use in ground beef products in 1994. Obviously "Pink Slime" was a bad descriptive term. The media jumped on it and slammed producers of LFTB for putting Pink Slime in America's food supply. The bogus misleading reports resulted in the removal of these products from the marketplace and the loss was equivalent to losing 1.2 million head of cattle. The market for these products is slowly coming back, but the damage has been done. It remains to be seen if the manufacturers will be able to collect their billion-dollar lawsuit against ABC.

The Pink Slime or more correctly LFTB episode is another illustration of how vulnerable and ill-prepared our industry is for misleading and shameful media coverage. It seems ironic and unfair that the beef industry is often targeted by these problems whereas the pork and poultry industry is not. Perhaps it is the price we pay for producing a high-dollar preferred product from animals that most of us love. Most of us and many consumers love cattle, and I doubt the same can be said for chickens and swine.

The years of 2011 and 2013 set all types of records in the beef cattle business as illustrated in Major New Millennium Events in Chapter 30. The record cattle prices are welcome, but we could do without the record feed prices. Obviously high fat cattle prices did not always mean profitable feeding programs. As long as there continues to be a 25–30% excess feedlot capacity, feedlot profitability will likely be elusive.

Regarding business, most of my time the past two years was occupied by land trades and cow trading. Some of this was for entertainment as I get easily bored if I'm not busy or sleeping. I sold the remainder of the California ranch plus the Oklahoma outfit. I hated to part with the California ranch, but it got very pricey. California bureaucracy is also aggravating, and more important I would see Caroline behind every rock and cow. Fifteen hundred miles is too far to travel to be melancholy and depressed.

–K–

Aging Well

K. S. Eng

Caroline was unique
Ageless you might say
If you knew her through the years
She grew more beautiful each day

She was like a good red wine
That peaked at an older age
Or like a good mystery book
That got better with each page

Or think of your favorite beef
If you'll let the carcass rest
Those extra days of aging
Gives you the very best

Maybe you enjoy a good cognac
From France where it's shipped
It is always most satisfying
When it's aged and slowly sipped

Sometimes she could be wild
You would never call her tame
And though she left us too soon
She left at the top of her game

One of the great memories on the Carrizo ranch was each Christmas when Dave and Diane Myklegard would come for three or four days. Sometimes Dr. Ray Cerniga and Chic would join us. The first time Ray came out, Dave gave him directions and told him Kenneth and Caroline lived quite a ways out in the "sticks." It might be a good idea to bring some groceries and wine as they don't have much. Ray pulled up in a car with the backseat full of groceries and wine and I said, "What's this all about?" Ray looked around and said, "It's about having a con man named Dave Myklegard as a friend."

Diane, Caroline, Dave and I would eat, drink wine, play pinochle and watch football games for three or four days straight. One holiday season we had a big storm that knocked out all the electricity. We played cards by candlelight and the light of the fireplace for two nights. Dave has more friends than any Norwegian I know, and their friendship has helped me make it through the night.

In 2011 I bought a couple of irrigated farms in Nebraska. I thought they were expensive when I bought them in the spring of 2011, but the price kept going up. Dinsdale called one day saying he heard I was buying Nebraska land.

I said, "I was but you Nebraska folks have priced this Texan out of the market."

Dinsdale answered, "That's because Nebraska people don't have as many girlfriends, so they've got more money to spend on land." Could have been true at one time, but not now.

Beginning in 2012 I've bought ranch-timber properties about ninety miles north of New Orleans in South Mississippi. I sold Nebraska farms to pay for the Mississippi property. We've harvested some timber, planted it to rye grass and converted it to cattle. I've got a lot to learn about the timber, but what I like is you always know where the timber's at, and usually there's a saw mill asking to buy. That beats my experience in the cattle business. We've got eighteen-plus miles of Pearl River frontage, six lakes and fifty-four inches of rain this year. For a fellow that's been in a drought all of his life, that's appealing. Another Mississippi attraction is our neighbors, the Harold, Ted and Roger Parker families. If you can't get excited with the Parkers around, you'd better check your pulse because you may be dead. When Harold is not busy with cattle, he's hunting, usually with his hounds. I believe he said he has sixty-nine treeing walker hounds. Whoever said you can't have too many hound dogs has never met Harold. About thirty years ago, I first met Harold in New Orleans and he, Dawn and Billie and I went to see a new cattle receiving operation that Oscar Black had built. I'd never met Oscar before and I believe he wore a silk shirt plus four or five gold chains around his neck. After we left, Harold asked if I thought I might do some business with Oscar and I said, "I doubt it. I've never had much luck doing business with a fellow who wears a gold chain and Oscar's got four of them."

Oscar was a good cattle trader—maybe too good—and he's spent some time in jail. Years ago Harold and I were dove hunting with Ed Bricker in Arizona when Harold got a call telling him he had better get back to Mississippi. They had found Oscar Black dead in a ditch. A Mississippi timber man didn't like how Oscar treated him on some cattle trades, and he hired a hit man to get even. Unfortunately the hit man was an undercover agent and they put out a news report on television that Oscar was dead. Harold said he was glad Oscar wasn't dead, but it sure ruined a good dove hunt.

I've said I never wanted to be dry again, and with the irrigated land in Nebraska and the Mississippi place with a fifty-four-inch annual rainfall average I was covered. I suppose the next thing will be hurricanes, but at least I'll be wet.

About a month before Caroline died and two years prior to my 75th birthday, Caroline and Annie started planning a surprise 75th birthday party for me. Annie went ahead with the plans and on March 3, 2012 she and approximately 150 friends and family

What Money Won't Buy

K. S. Eng

There are those who worship money
But when the final bell is rung
They are usually disappointed
At how little they have won

Making money can be a thrill
But they often come to see
What it bought was disappointing
Because the best they got was free

We all would like more money
But regardless of your wealth
Money won't buy what's most precious
Peace of mind and good health

Money won't buy you true friends
Special friends who care the most
Money seldom buys true love
Although some nights it may come close

Money won't buy a soft summer rain
Or the blessings you like to share
It won't buy a sunset or a rainbow
Or a blue moon that's so rare

What can you buy with money?
More lawyers, accountants and greed
Money will buy more play things
But how many toys do you need

But if your money buys you freedom
To do what you love to do
Consider yourself more than lucky
You're one of the chosen few

sure surprised me. Because Caroline and I travelled and lived in so many different locations it was hard to find a central location but they chose San Antonio. I never had a clue until I walked through the auditorium door. Everyone said they never thought they'd pull it off without me knowing but I told them it's not hard to fool an old man. About six months prior to the party, Annie informed me she was going to San Francisco on vacation the first week of March and be sure to put that on my calendar. Laura and Julie also called and said they wanted to be in San Antonio on my 75th birthday so not to make any plans. I called Annie several times on her mobile while she was "in San Francisco" and asked her about her trip, her hotel, and so forth. She said San Francisco was lovely and it was a great boutique hotel, but the guys were a little too cute. All of that time, she was only three or four blocks away with several friends preparing for my party. It was a great party and best of all, it made me appreciate how fortunate I am to have so many good friends. Walt Davis from Hereford, Texas was in charge of the steaks, and it doesn't get any better than that. Annie put together a memory book for me and many of the birthday party pictures and memorabilia are on the following pages.

Kenneth and Dr. Russell Cross, Texas A&M

Vi Reeves with Scott and Lynette Shrauner

Douglas (Caroline's son) and his wife, Debbie.

Baxter and Violette

HAPPY BIRTHDAY, KENNY!!!

You have come a long ways during the past **75 YEARS!!!**. We are all very proud of your accomplishments.

It was during the late 1950's and early 1960's that I had the opportunity to recognize your talents and wisdom. It was a privilege to have you as a graduate student and as your major advisor to earn your M.S. degree at University of Nebraska. I also found in my records that you were in my Feeds and Feeding class of 95 students in 1956-1957 with your home town listed as Newman Grove, Nebraska. You came close to receive the highest grade in that class. You probably could have received the top grade in that class if I was able to READ your answers. I probably should have given you credit for your LEFT HAND writing (only kidding).

Besides being a prominent livestock consultant in the United States for many years your talents as a POET have been admirable. You have been kind to send me copies of your calendars each year and they have been appreciated.

Caroline will be missed at your 75th birthday but we will all be thinking of her. I wish you the best in years ahead and hope you have many more HAPPY BIRTHDAYS.

Sincerely yours,

John

John K. Matsushima

March, 2012

Happy Birthday, Kenneth Eng

Once upon a time, a young girl, namely me, Ginger, moved to Newman Grove, NE to be a bookkeeper at Wobig Implement Co. It was a different little town then I was used to. No one knew how to dance, and there wasn't much action, except on Poker Hill! I started to make new friends, and once while a group was standing around talking, I saw this neat guy, and asked Babe Anderson, "who is that cute guy?" She said, "that's Kenny Eng, but he's taken by Jeanie Farrar." "Shoot", I said, but as time went on we got to know each other. Time passes and everyone was graduating and going on with their lives, and I decided to go back to Knox County as I had gotten a job in the Courthouse.

When reading the Norfolk paper, I found out Kenneth Eng was boxing in the Golden Gloves. I knew he had such a strong will and desire to do whatever he set out to do, that me and my boyfriend went to watch him box, and he did so good, I was quite proud that I knew him.

Disaster struck, as a car my boyfriend was driving hit head on with another car on a country road detour. It threw me into the windshield, breaking the glass and much of my face. It was tough times emotionally and physically. Ken heard about the accident and got in touch asking me to go out. Even though, I was getting married, he never knew how much that me to me, as I was fighting some esteem problems, due to facial surgeries and scars. I had to write and tell him I was getting married. He later told me he kept my "Dear John" letter and still has it somewhere. What a guy!!

As the years passed, Ken still tried to keep in touch. He would stop at the Courthouse, where I now was Knox County Clerk. It was always good to see him and catch up on our old Newman Grove friends. Time went on and we both lost our spouses.

Later he purchased land in the area and left a calendar for me at my local bank. It was filled with his poems, which gave me a look into his life. I called and thanked him for the calendar and remembering me. We visit now and then sharing our active lives. But now "dear old friend" we've both made it to 75, and even though we're on the down side of advantage, we still have lots to offer.

It's been a pleasure knowing you since we were 16. We've done a lot of living and I look forward to much more as I know you do too.

As Ever,

Virginia

HAPPY 75TH BIRTHDAY TO KEN

We were asked to write some story or experience of what we remember about you.

Well, I had to think for awhile to see what or which one I could write about. First off, I cannot believe that you are celebrating your 75th birthday this year. You do have lots of company this year for that milestone.

Do you remember when we were accepted into the Alumni Association of Newman Grove High School? I remember we were looking over the crowd of Alumni's and of course we were looking at the class of 50 years and more, thinking "boy, are they old" and here we are even older than that.. Where has the time gone?

First and foremost what comes to my mind is that you were the life of the class room. The one thing that I think of most is how you would come into Mrs. Hanson's history class always laughing. You were always doing that and usually a little late. I think that Mrs Hanson did not know what to do with you. We would always have to laugh a little as well because you seemed to not really care. I think it drove her crazy!!!! I do not think that any of the rest of us could have gotten away with what you did.

You must have been absorbing more of the intellect side of high school than what we observed. You have done and accomplished so much since high school in your business life and your creativity that you have always shown us. Although as I am writing this I realize you were always looking for creative ways to have a good time.

I do want to thank you for all of the memories and the good times we have shared since high school, at the alumni banquets, getting together some times when you come to Omaha. It truly is a blessing to have been a part of your life in Newman Grove and elsewhere.

Keep in touch and share your poems with us like you have in the past. We really enjoy getting those books and enjoying your humor. Also, the poems sharing your thoughts and how much you miss Carolyn.

Have a great birthday and sorry we could not make it to your birthday bash!!!!

On The Trail With Ken Eng

Remember the early meetings in Casa Grande/Stays at the condo in Phoenix/Visits to the condo in La Jolla/Catching up in Idaho/Late night in Amarillo/Great barbecues at the New Mexico Ranch/Profound discussions at the Carizzo Ranch "oiled" into the wee hours with generous pourings of wine by Caroline/More sedate times in San Antonio.
Add to this business ventures, fishing trips to Alaska, dove shoots at Brickers feedlot, Plains Nutrition Council and Southwest Nutrition Conference meetings, Feedlot cattle and National Cattle Association meetings in various cities and more.

It's been a great trip, the adventure continues.
Great memories to take into our upcoming Golden Years, which are still a ways off.

Congratulations and Best Wishes,
Ann and Fred Klatte

Ann + Fred Klatte

Birthday party 2012

The persistent trends of 2010, '11 and '12 were the relentless decrease in cow numbers and increase in land values. Also drought was the problem in many areas of the country, feed prices reached all-time highs and feedlot profits were illusive or non-existent. Cow and calf prices were strong and combined with accelerating land prices resulted in the largest transfer of wealth the cattle industry has ever experienced. It's not surprising that some ranchers and cattlemen elected to take the money and run, resulting in more cow liquidation than one would have expected in profitable times.

Although profits were great for the land and cow owner, it was obvious profitability needed to return to the feedlot industry if the industry as we know it was to survive. Another reason for purchasing the Mississippi properties is I've always had a theory that if you have a chance to buy productive land versus pretty land, buy the pretty. If you buy productive country, someone may come along and make other land even more productive., but if you buy the pretty land, it will always be pretty. Someday someone may come along and want to pay more for it because of its beauty. The California and New Mexico ranches were in the pretty category.

The Nebraska land was in the productive category, and I was just fortunate to buy those farms on the way up and rode the rising land price tide to some pretty high levels. They say beauty is in the eye of the beholder and I suppose that will always be true. To me the Mississippi places are flat out beautiful. They're not what you would call extremely productive, but with the combination of timber, grazing, hunting leases, and so forth, the cash flow is pretty good. Look at the following pictures of the South Mississippi place and see if you don't agree, it's a Southern version of "God's country" and with people to match. At my age, this is probably my last stop. Maybe I'll leave the kids and grandkids places that are both pretty and profitable.

The Mississippi properties have far more improvements than I need but you can't buy someone's land and timber and ask them to keep the improvements. Among other items, there are eight houses, a lodge with a large kitchen that will seat and serve

Mississippi House

approximately 250 people plus lakes and Pearl River frontage. The main house is approximately 10,000 square feet, and for me it's an embarrassment of riches. It's not something a seventy-seven-year-old widower needs. Annie and the girls believe they can keep the facilities occupied with seminars, banquets, wedding, parties, and more. I hope so, because the overhead is pretty steep. Bottom line: I've got the land, water and timber of my dreams, and I guess I'll have to live with the overhead.

Cypress Trees

Mississippi Lake

Eulogy by Baxter Black for

CAROLINE ENG Jul 3, 2010 p1

Most of you here know Caroline and Ken as personal friends or as family members. I know them on the world stage where they were stars. I met Ken in 1970, a renowned consulting nutritionist. I was a rookie feedlot veterinarian, wet behind the ears. After he met Caroline they ascended to royalty in our little world. So when I tell you I am honored to be invited to be a part of today, you know it is coming from a humble heart.

I got the call. Mark Adams said Kenny was trying to get a hold of me. I hung up and called him. His wife died. An immediate flashback put me in a hotel room five months ago at the NCBA meeting. Caroline and Kenneth Eng, the ever present blue-ribbon hosts, celebrating with their friends. Musicians like Dr. Cerniga and my wife, Cindylou were playing country classics, Kenneth tapping his toe and Caroline by his side, watching the crowd to see if everybody was having a good time.

We were! As LBJ would say, "They treated us like a Kennedy!"

Fast forward to the phone call. Kenneth explaining the circumstances, trying to maintain control as grief squeezed his heart like a vise. He took a breath and concluded, "We were both strong believers even though we didn't go to church much. Caroline was raised in a strong Baptist upbringing."

His proclamation of faith immediately set my heart at ease. I live with the dread of attending the funerals of friends who deny the existence of God. If invited to speak, what could I possibly say?

We live in a world where our faith is often kept in a quiet place. It surfaces under stressful conditions. I remember the anguished voices on television of people running from the exploding carnage of the Twin Towers. Over and over, shouting and crying "Oh, my God, Oh, my God. Oh, my God!"

I remember thinking then, that's with a capital G.

It has been said, "You never know how much you need Jesus, until He is all you need."

P2

Losing a wife is a sterling, maybe the ultimate, test of one's faith. To hold the unquestionable belief that Heaven, however unfathomable it may be, is real and that you will be together again for eternity as promised by God Himself.

It is one of those times in our life where you have to completely let go and trust that God will catch you. And know that however lonely and grief stricken you feel, this is not the end of your relationship with the ones in Christ that you love. LO, I AM WITH YOU ALWAYS EVEN UNTO THE END OF THE WORLD. Matt. 28:20 God's world, not yours.

Ken and Caroline have God on their side. She has already walked through the valley of the shadow of death, fearing no evil. She has done it. She's already there. She knows she'll see Kenny, and all of us again. Including you. Everyone has an invitation. Simply believing gives you a pass.

I guess that means she is here this morning, too, watching us all, who came to pay our respects. She understands how much it hurts us now, she can feel our pain, but time for her now has no meaning. In the blink of a Heavenly eye the surly bonds of Earth will release our battered souls to join her and she is getting the party ready. The Heavenly hosts has agreed to play music for us. She has two fish, five loaves of bread and three bone-in Rib Eyes ready to be blessed and feed us all.

It will be a party where the packer will lie down with the feeder, the banker will wash the rancher's feet, and the market will always be up. Peta and the EPA will be doing eternal Hail Marys, paying penance for their mistreatment of those hard working cattle people whose lives they have made miserable.

There are not many essential jobs on earth. I mean, cowboy poet? Maybe that's why I've always had to run a few cattle in my life to make a legitimate contribution. That is the reality of real life. Feeding the teeming masses, be it with wheat, meat, fish, venison, tortillas, or pinto beans, is one of the essential few jobs on earth. Caroline and Ken have helped put a lot of food on a lot of tables. They did it with the help of many friends, family, employees, partners, and guardian angels, a lot of whom are here this morning. They are still doing it.

(I read three of Kenneth's poems about Caroline. She was his inspiration for so many of them) THE WINNING PAIR, THE SMALLER THE MEANER, WE'RE STILL A PAIR

Aunt Effie and Uncle Leonard, of my Oklahoma lineage, were farmers. They were peach cobbler-mule drawn hay rakes-fiddle music-church on Sunday-open doors and caring hearts.

Married 70 years, they were each other's beating heart and sweet dream. Aunt Effie said, every night before they went to sleep they would tell each other how much they loved each other. I imagined she would look into his eyes and say, "When I slide in under the covers and my leg touches yours, it's like petting the cat till sparks fly off its tail!"

Then he would reply, "If loving you was a handicap, I'd always have a place to park."

Many of us can relate to losing a loved one. A month ago, nearly to the day, we scattered my mother's ashes on my father's grave. Grandpa Tommy's, too. She was 91. She had buried two husbands. She was ready. What helped sustain her was the knowledge that, as she put, "I knew what I had when I had it, and I knew what I lost when I lost it."

Caroline's life on Earth has ended but her love for us, and ours for her, has not. It will go on forever.. It is the greatest gift that one can give or receive, returned love. To love someone who loves you back. Ken and Caroline had it and were blessed by having each other. Sometimes we just get lucky.

yer friend, baxter 2010

When Life Doesn't Go As Planned (Things I Learned from Mother After Her Mind Was Gone)

K. S. Eng

I learned a mind is precious
And doesn't always work as planned
I learned there is a God
But He's not easy to understand

I learned the values of old friends
Who would visit her each day
And the goodness of a country
minister
Who would hold her hand and pray

I learned of dedicated caregivers
Who cared for her every need
Who had Patience and dedication
And would follow her or lead

I learned to cherish the good times
It's one of nature's healing ways
For if we dwelled on the bad
There'd be no more good days

I learned to enjoy my friends
Because life is built on shifting sand
And remember to say "I love you"
In case life doesn't go as planned

-CHAPTER 29-

My Business Game Plan—What Game Plan?

Those who know me well also know that when it comes to business I often change game plans. I subscribe to the Willie Nelson and Kinky Freidman School of Business which says if your horse stumbles or dies, dismount. I believe the most important rules in business are flexibility and an exit strategy. Some say whenever a door is closed, another opens. My theory is don't hesitate to open a new door before the old door closes. Because of this philosophy I've never really had a long-range business plan.

I'm occasionally asked to speak to student groups and others about my career. I'm sometimes introduced as a smart fellow and risk-taker who worked hard and as a result became successful. I don't believe this is entirely correct, and it's not false modesty to say I don't believe I'm smarter than many others. However, I have had more than my share of good luck and have had good timing. I believe everyone occasionally has good luck although some more than others. However, it's timing and the willingness to take advantage of the good luck that separates the men from the boys. If you have an unusual opportunity due to good luck and you don't act on it, years later you'll look back on it and say, "would've, should've, could've," and you're no better off than the fellow with no luck at all.

In life, business and love, good timing is crucial, but occasionally you will make a move too early that doesn't work as you hoped. It helps to have an attitude where losing money or a little personal embarrassment won't ruin your day. Personally, money, as such, has never been really important to me, and neither has embarrassment or what others might think. Bob Rebholtz used to say, “Kenneth, you’re not hard to embarrass, you’re impossible to embarrass.”

While I don’t really have a long-range game plan, I do have several rules for a successful game plan. They are as follows:

1. When you can’t differentiate between your work and your play, you know you are successful. Do what you like and like what you do.
2. Never pass up a chance to see something new, and if possible, see it firsthand.
3 Learn from your mistakes, and try not to repeat them.
4. Don’t hang around boring, negative people. They will sap your energy.
5. Take care of yourself, and don’t abuse your body (unless it’s a really good party).
6. Avoid dishonest people, and if it seems too good to be true, it probably is.
7. Try not to worry about things you can’t change, and don’t sweat the small stuff.
8. Jealousy is the single most destructive personal trait. Avoid jealousy and rejoice at your friend’s good fortune.
9. Recognize good luck when it occurs and take advantage of it. This often translates into good timing.
10. Be flexible and have a business exit strategy.
11. RETURN YOUR PHONE CALLS.

The Game Plan of a Fool

K. S. Eng

I called an old friend today
One I hadn't talked to in years
His secretary said "he died Saturday"
I couldn't stop the tears

She said "he spoke of you often
And the good times you both enjoyed
He wished you both weren't so busy
That happens when you're over
employed"

When revisiting lifes mistakes
This is lifes saddest line
"Guess I was always too busy
Sorry I never found the time"

I didn't always say thank you
Or ask if you were feeling fine
I thought my friends should know
I was always short of time

So take time to smell the roses
Live life by the golden rule
And remember all work and no play
Is the game plan of a fool

Retirement Plan

K. S. Eng

When they ask us about retirement
The wife will always allow
"Took me forever to get him trained
Can't let him quit just now"

We've always invested in cattle
And lot's of money has been lost
Figure I need 20 more years
To get back to our original cost

Something new is always happening
It's exciting through "thick & thin"
Why should I worry about retiring?
My bankers will tell me when

We want to do more traveling
And wherever we may roam
I can always do a little business
Because I've got my cellular phone

So many old friends I want to see
That'll take a lot of time
So many old haunts I want to revisit
And "return to the scene of the crime"

We've got a little money saved
That my wife is dying to spend
And if that money should run out
The banks are still willing to lend

Why would you want to retire?
If there's still "want to" in your bones
Why not plan to leave here broke
And "maxed out" on all of your loans

When this game is over
They had better bury me deep
I'm having too much fun to quit
I just need a little more sleep

-CHAPTER 30-

Mid-2012—Mid-2014
(My Post-50-Years Era and What a Ride)

Originally I planned to complete this book in June 2012 which would mark my fifty years in the industry. However, as events unfolded at that time, it appeared the remainder of 2012 and a good portion of 2013 could be catastrophic and a game changer—especially for the feedlot industry. As much as I would have liked to ignore these developments, it would have made the historical perspective of this book incomplete. Furthermore, it appeared the industry could turn around in late 2013 and 2014, and if possible, I wanted to end the book on a positive note. I had no idea how correct my projections would be, and again lady luck was by my side. Finally, I was beginning to recover from my depression following Caroline's death, and reasoned that my attitude had to improve or I'd be pushing up daisies.

No book on the history of this industry would be complete without mentioning some of the frauds and scams that have all too frequently occurred. On the surface, this seems unusual because we are an industry that prides itself on the value of a man's word and a shake of the hand. However, we're also an industry that respects people's privacy and it's often been said, "Never ask a man about the size of his herd or his poke unless he volunteers the information." In some instances I believe frauds and scams go undetected or unreported longer than expected because some who are aware of them don't volunteer information unless asked.

I recall when I was first working in Idaho in the early '70s, a couple of cattle traders were selling Monfort large numbers of feeder cattle. It was widely suspected by cattlemen in that area that these people were padding weights and prices at Monfort's expense. However, it was several years before the Monfort organization became aware of the scam and put an end to it. An interesting footnote is that one of the perpetrators supposedly died in a private plane crash shortly thereafter. However, there have been persistent rumors that it wasn't him in the airplane with some claiming to have seen him only a year or two ago.

The biggest cattle fraud in history came to a close in 2004. It involved two cattle brokers, George Young and Kathleen McDonald, from the Kansas City area. The losses experienced by banks and investors was said to total $183,000,000. There were approximately one hundred thirty banks, individuals and businesses victimized. This scam resembled a Ponzi scheme where fictitious cattle were sold to the investor and profits were guaranteed. At the end, when they inventoried actual cattle numbers I believe there was only 14,000 head, and these had been counted over and over again in various investor accounts. Some feedlots and cattlemen not involved in the scam suspected the operation was fraudulent because there was no way they could guarantee the profits they claimed. A client of mine in Nebraska had a neighbor who had invested several million dollars with George Young. I asked if he'd told his neighbor it was a questionable investment. His reply was, "No, he's a grown man and he's never asked my opinion."

Most recently we had the Eastern Livestock bankruptcy and conviction of Tommy Gibson the President for fraudulent cattle trades. Losses at Eastern Livestock may exceed those in the George Young fiasco, and it was a devastating event for many. Obviously, it was devastating to cattlemen who lost money because they were not paid for their cattle, and for those who bought cattle they didn't receive. It sent shockwaves throughout the industry because

Eastern Livestock was by far the largest industry cattle dealer and Tommy Gibson was well known and respected by most. Caroline and I sold a lot of cattle to Tommy in the '90s—some on the shake of a hand and we never had a bad deal. The full story concerning the Eastern Livestock collapse may never be known, but I suspect there are two sides. Some people who Tommy traded with did not treat him right, and that might have included some banks.

There are numerous other crooks and con men who have defrauded cattle people and banks. The list is too long to detail but it's fascinating how many are repeat offenders. If investors would do their homework, they might not get scammed in the first place. My rule when considering an investment with someone we do not know is that three phone calls will probably give you the real skinny. The man you call first may tell you he doesn't know the individual, but suggests you call number two who does know him. You call number two and he may reply, "I don't know him well, but call number three and he can give you his history." This strategy has saved me a lot of grief and money in my lifetime. One final thought, in bankruptcies, many people come up short of money, but the bankruptcy attorneys always get paid.

Although not a cattle business scam, perhaps the most colorful con man in agriculture was Billie Sol Estes who died in 2013. He built a fortune on nonexistent fertilizer tanks, fake mortgages, bogus cotton acreage, etc. in the 1960s. Farmers and bankers in four states were bamboozled before Billie Sol Estes went to prison. There were also a string of strange and suspect suicides, including an investigator who reportedly shot himself five times with a bolt action rifle. Billie Sol Estes was from Abilene, Texas, a bible thumper and a friend of many powerful politicians including Lyndon Johnson. Although never proven, some of his statements included allegations he had given Johnson millions of dollars, and Johnson had ordered seven killings to be disguised as suicides and accidents to cover his connections in the frauds. In total, there were said to be seven suspicious deaths associated with the Billie Sol Estes saga.

Perhaps the most audacious and brazen scam I'm aware of in the cattle business involved a couple of Texas-New Mexico border traders approximately twenty-five years ago. The perpetrator sold a large string of Mexican calves to another trader, and then offered to find grazing for the calves for the next four or five months at a set fee on a per head per month basis. The trade was completed, the seller got his money plus a grazing fee to be paid each month. He then resold many of the cattle to other unsuspecting buyers, but continued to bill for grazing for the next several months. In the end, the original buyer not only lost most of his cattle, but he also paid for several months grazing on the non-existent cattle. Talk about the worst of both worlds! Larry Titsworth, a border trader friend of mine, commented that even if the Lord himself was pasturing his cattle, he'd check them at least once a month. If the con man involved in this trade whose bad habits were widely known was grazing the cattle, he would keep an armed guard at the gate 24/7.

True to my gut feeling, cattle prices in late 2012 and the first nine months of 2013 were a disaster for the feedlot industry and disappointing for the stocker people. Many feedlot cattle were placed with projected $140+ break-evens, and market bulls assumed that $140+ fat cattle were a sure bet. Once again the bulls were wrong, and in reality the market peaked in the $128 - $132 range. That's not bad from a historic standpoint, but it was well short of the record break-evens. Some losses reached $500 per head which would rival the worst we saw in 1974. I thought we would never see another wreck of that magnitude, but late 2012 and '13 almost proved me wrong. The feedlot situation was complicated by a change in packer policy that drastically changed the basis on calf-fed Holsteins in some areas. Under the new basis bids, previously uncontracted feedlot Holsteins lost as much as $500 per head. One feedlot Manager/Owner told me he'd rather have a social disease than a pen of uncontracted fat Holsteins.

A piece of good news was that drought conditions prevalent over much the United States in 2012 began to moderate. Rainfall was good in most areas east of the 98th meridian which included Central and Eastern Kansas and Nebraska. Moisture conditions were also very good in the Southeast from East Texas through Florida. As a result, grain and hay prices decreased significantly. However, the droughts persisted from the Texas and Oklahoma Panhandle and westward all the way to the Pacific Ocean. The past three years have been the driest on record for the Texas Panhandle.

The calf and yearling prices were also disappointing the first half of 2013 and were $20 to $30/cwt. under predictions. Packer cow prices while good were also under projections. Even so, cow slaughter and liquidation continued to increase.

Many in our industry began questioning our leadership. The relentless theme of industry consolidation and the "big getting bigger," provides a level of discomfort and skcpticism among many cattle people. Whatever the game plan may be in our feedlot industry, it was obviously flawed. While the outlook for the ranch and cow industry is positive, herd expansion is slow to start. Without meaningful cowherd expansion was slow to start, our industry will continue to downsize and become less relevant. This is not the picture most visualized at the beginning of the new millennium.

The 2013 cow-calf business was profitable again even with higher feed costs early in the year. This should lead to cowherd expansion later in 2014. That would seem a good bet since the cow-calf business has been profitable for the past ten years. However, concentration in the feedlot and packer segments makes many cow men nervous. The thought of losing market leverage is worrisome.

Our market share compared to poultry is also declining. It is estimated that our per capita beef supply in 2013 dropped approximately one pound compared to an increase of 1.2 pounds for poultry. Furthermore, lower grain prices will help the poultry and pork industry more than the beef industry. Poultry production is far ahead of both groups.

Some negative people believe we're headed in the same direction as the sheep industry, but I doubt this. Like many, I seriously doubt that consolidation and "bigger is better" is working. Perhaps it never did. Currently, 90% of the cow operations in the United States contain 100 cows or less and 2/3s of the calves are produced by operations containing 1000 cows or less. I suspect when one considers the increased value of their land plus cattle, most producers have done well. Moreover, small ranches have improved cattle productivity and quality, and are doing a pretty good job for the industry.

While the cow-calf business has been profitable the past ten years, the feedlot segment has been profitable only in 2003 and late 2013 and 2014 with short-term profitability in a few other years. Consider the marketing strategies in the different segments of the cattle industry. In general, the cow, calf and yearling market is determined at livestock and video auctions or by private treaty. On the other hand, most feedlot cattle are sold on a "formula and contract basis" with less than 25% sold on the cash market. This leads one to conclude a major flaw in the feedlot industry is the marketing structure, and increased size and concentration hasn't solved this problem.

With this economic background, it is not surprising that drug companies appear to be spending relatively less on beef cattle research and development. There are far more family dogs and cats than cattle, and many drug companies are emphasizing companion animal research drug development. No doubt the withdrawal of Zilmax from the market will also have a chilling effect on future R & D.

The following list of new products for the beef cattle industry introduced since 2000 is painfully short—especially if the introduction of generic products are not included:

New Products Cleared >2000

2000—Rev IS & H

2003—Synovex Choice

2004—Optiflex

2006—Zilmax

2007—Revalor XS

Generic Approvals Since 2000

2001—T.E. Implants (Trenbolone Estrogen)

2008—M.G.A

2012—Tylan

2013—Optaflexx

2015—Monensin?

In August 2013, because of real or perceived animal welfare problems related to Zilmax, Zilmax was taken off the market. The industry is divided on the issue. This decision is further complicated because Zilmax was removed from the market, carcass grades have improved and fat cattle prices have increased.

Animal health problems in both the swine and cattle industry seem to be increasing the past several years. Porcine Epidemic Diarrhea Virus (PEDV) is causing high mortality rates in baby pigs in approximately twenty states and Canada. Had this virus not occurred, hog production in the U.S. in 2013 and 2014 would have been substantially greater.

There has also been a substantial increase in disease problems in calves in some areas, especially those of sale barn origin. Many stocker and feedlot operations which rely on sale barn calves now routinely pencil an 8% to 10% death loss and a \$50-\$100 per head drug bill. Long-term this is unsustainable. Even with preconditioned calves such as the Vac 45 program, many feedlots are requesting the post-weaning period be extended to sixty days to reduce health problems. The feedlot and veterinarian community has not found the answers and many are critical of the drug industry believing existing drugs and medications are becoming less effective. Hopefully we'll see improvement in this area, but we should not expect drugs and vaccines alone to provide the only answer. We have noted in our confinement cow programs where we do a good job of feeding both the cow and the calf, weaning stress is greatly reduced, and in general, calf health is very good. This suggests that both pre- and post-natal nutrition is important in calf health.

With certain classes of calves, the current animal health situation is an impending "train wreck." The possible good news is that the extra high calf death loss means the projected "wall of calves" coming at our fat cattle industry this spring and summer may be smaller than anticipated. The high feedlot placements we've seen for the past several months are based on feedlot purchases and not on purchases less death loss. Health issues should be addressed—and the sooner the better.

There has been more discussion about reducing or eliminating the corn-ethanol mandate and ethanol use in gasoline has declined. However, demand for ethanol has actually increased partially due to increased export demand. Also, even with lower corn prices and increased distillers' production, corn distillers' prices have been up and down depending on export demand. The supply of gluten, largely a by-product of corn fructose production, has decreased. Some call this the "fat girl syndrome" because of a large reduction in soda consumption that means less fructose is required.

To some extent, the nutrition and feedlot industry is responsible because when gluten and distiller supplies were abundant, we drove some pretty hard bargains and used high levels of these by-products in our feedlot programs. Nutritionists and feedlots who were on the "cutting edge" of technology frequently fed rations containing high distillers and gluten combinations (DMB) because it was a bargain and gave us the least cost gains. As more feedlots and other animal industry groups learn to use these by-products and as international demand increased, by-product bargains have diminished.

In retrospect, it might have been better had we not bargained so hard on corn distillers prices (especially the wet material). However, feed users felt little sympathy for the ethanol producers who were receiving government subsidies and pushing corn prices up. Most feeders looked at feed by-products as a way to get even.

Some believe grain ethanol production will decrease but cellulosic ethanol production may increase. While that's possible, it would come at a cost. Contrary to what some think, roughages and cellulose by-products from which cellulosic ethanol would be produced are not inexpensive. Based on cost per unit of nutrient density, roughages are often the highest priced ingredients in our rations. Some of the cellulosic ethanol production estimates would utilize almost all of the available fiber now used by livestock. Technology in the cellulosic ethanol production appears to be improving, but be careful what you wish for.

The Government Biofuel programs will go down in history as one of the "game changing" events in agriculture. Like many other government programs it benefited some and was unfair to others. It obviously favored Midwest farmers and cattlemen as compared to feedlots in the South Plains and the West Coast. The resulting high grain and high land prices added enormous wealth to the Midwest section of the country. Furthermore, occasional "bargain basement" prices on the wet byproducts benefited the Midwest feeder as compared to those in more distant areas. The biofuel programs were a windfall to some and unfair to others. Unfortunately, life is sometimes unfair and we had little choice other than to deal with it.

Even with less ethanol used in gasoline, current ethanol production is record high and may be here to stay. An unexpected consequence of the increased corn demand was a lot of "transition land" in Nebraska, South Dakota, etc. was converted from pastures to crops. Just as with the Homestead Act in 1862 the end result was prairie and grass plowed under and converted to crops. Results included less pasture land and higher pasture lease land prices. Time will tell if these changes are permanent and part of a longer term trend.

The late 2013-14 winter has been record cold. One hopes this would quiet the global warming lobby. However, to many global warming proponents, facts are irrelevant and have little place in their doctrine. When I was on the "dog and pony show" speaking circuit in the '70s, climate change was of major interest. Then the climate change worry was global cooling and some projected that by 1980, no corn would be grown north of Ohio. We had several good speakers on this subject who believed global climate was to a large extent determined by volcanic activity. If there was widespread volcanic activity, the global climate would be wet and cool, and if there was not, the opposite would be true.

Obviously, long-term global climate cooling never occurred, but many of the speakers agreed that human activity had minimal impact on global climate change. That made sense to me then and it still does. Many currently predicting global warming appear arrogant and short sighted. Some global warming proponents are definitely not our friends, and their real agenda is anti-capitalist and anti-animal agriculture. Recently, there has been lip service paid to new studies suggesting emissions of methane from fossil fuel and from cattle flatulence has become a major source of global warming in the Great Plains areas. One wonders if they considered that 150 years ago the buffalo population in those areas may have been twice the current cattle population. Furthermore, since the buffalo's diet was almost totally forage, the methane production would have been larger than with current livestock diets. The cattle flatulence issue may be a "backdoor" attempt by some to regulate our industry.

On September 13th and 14th, 2013 we held our first Kenneth & Caroline McDonald Eng Foundation cow-efficiency symposium in Lincoln, Nebraska. The University of Nebraska Animal Science Department hosted the meetings. They did a great job. Oklahoma State and Texas A&M assisted. We had over 200 people registered from fourteen states and received good reviews. For those of you who were unable to attend, the Proceedings are available for $50 each. The 2014 Symposium is scheduled for September 18th and 19th at the Embassy Suites-Riverwalk in San Antonio with Texas A&M University hosting and Oklahoma State University and University of Nebraska assisting. The Foundation will continue to sponsor research and symposiums in this general area of cow production. It is funded in memory of Caroline and her love for life, cattle and cattle people. The following is a paper I presented at the symposium on confined cow production. It has been updated to cover year-end 2013 market data.

The interest in confinement or semi-confinement cow production continues to increase. Ron Crocker, near Mason, Tex-

MAJOR NEW MILLENNIUM EVENTS

(If you are bored and not confused you're not paying attention)

2000 – 01/31/14
Does Semi-Confined Beef Cow Production Have a Place

K. S Eng, Ph.D.
Eng (-K- & -C-) Ranches, Mississippi, Nebraska and Texas

Major Cattle Industry Events (Short Version)

Past 13 Years

1. Relentless beef cow liquidation
2. Relentless land value increase
3. Severe drought first in Southern Plains and spreading to Central Plains and Southwest
4. Cows and offspring are getting larger, gaining faster with excellent carcass traits. But, how about efficiency (82% average weaned calf crop?)

January 1 Cow Numbers, Million Head

Year	Beef	Dairy	Total
1950	15.95	22.00	37.95
1960	25.68	17.65	43.33
1970	36.39	12.09	48.78
1975*	45.71	11.22	56.93
1980	37.11	10.76	47.47
1990	32.43	10.02	42.47
2000	33.57	9.19	42.76
2010	31.38	9.08	40.05
2011	30.86	9.15	40.01
2012	30.20	9.20	39.40
2013	29.30	9.20	38.50
2014	29.00	9.20	38.30

* A wreck

Average Land Price $/Acre

Year		Price
2000		1,090
2001		1,150
2002		1,210
2003		1,270
2004		1,360
2005		1,650
2006		1,900
2007		2,160
2008		2,350
2009		2,400
2010		2,500
2011*	Crop land	3,020
	Pasture land	1,090
2012	Crop land	3,550
	Pasture land	1,150
2013	Crop land	4,000
	Pasture land	1,200

*Individual states much greater increase, e.g. Nebraska land prices up 36% in 2011

Cattle Fax
Cattle and Corn Prices

	10/26/01	10/14/11	7/20/12	7/12/13	1/31/14
Slaughter Steer (CWT)	$64.97	$119.24	$113.00	$119.50	$149.00
550 lb Calf (CWT)	$91.34	$145.21	$149.00	$158.00	$204.00
750 lb Yearling (CWT)	$86.05	$135.15	$133.00	$143.00	$170.00
Utility Cow (CWT)	$39.12	$65.67	$77.00	$80.00	$89.00
Bred Cow	$800.00?	$1200.00?	$1,300.00?	$1,400.00?	$2,000.00+
Omaha Corn (bu)	$1.80	$6.28	$8.12	$7.14	4.26
May Feeders Future	$75.00	$148.47	$150.30	$150.12 (August)	170.07

3 Years of Records, 2011 – January 2013

1. More of the same
2. Record drought in South Plains (Texas, Okla. & N. Mex.) and then in Central Plains and Southwest
3. Record corn prices and 35% decline by January 2014
4. $150.00 top Fat Cattle
5. $220.00 Calves
6. $170.00 Yearlings
7. $95.00 Packer Cows
8. $105.00 Packer Bulls
9. $1,400 - $2,000.00 Bred Cows
10. $1,800 - $2,200.00 Bred Heifers

-K-

Major Cattle Industry Events

Net Results

1. Massive Cow Liquidation in TX, OK & NM in 2011 and in Central Plains in 2012
2. Smallest Beef Cow Herd in over 50 years
3. Continued Excess Feedlot Capacity
4. Record High Cattle Prices, especially Calves and Yearlings
5. Transfer of Wealth to former "Have not's"
6. Semi-Confinement Cow Production

   A necessity in some areas and an opportunity in other areas

Possible Short & Long Term Repercussions

Let's hope it doesn't occur but,

1. According to the Texas State Climatologist Dr. John Nielson-Gammon

"Another 5 – 10 More Years of Drought is a Strong Possibility"

–K–

Possible Short & Long Term Repercussions (cont.)

2. The 12 month period beginning September 2010 was the hottest and driest on record in Texas

   a) Resulted in Record Cow movement and Liquidation

   b) Feed Shortages and Record Feed Prices

   c) At very least it has been a Short Term "Game Changer" not only for Texas but the entire South Plains and in 2012 drought moved North to Central Plains (KS, NE, CO & WY)

   d) Should the Drought persist for several years it will change the face of the entire Beef Industry, 25% of the Beef Cows reside in Texas and Oklahoma

   e) It is difficult to feed your way out of a drought with high priced hay

3. Some Good News

a) Good rains East of 98th Meridian from Texas to Canada in 2013 but West Texas, NM, CO, AZ and CA remain extremely dry.

b) However, even with more rain Cow Liquidation is continuing.

c) When repopulation in Texas is possible, other species such as Sheep, Goats, Deer etc. may be considered. However predators are a major problem.

–K–

Potential Advantages of Semi-Confinement Cow Systems

1. **Reduced Cow Unit Carrying Capacity Cost**
2. **Reduced Cow Energy Requirements**
   - **a) Reduced Programmed Intake Increases Digestibility**
   - **b) Reduced Movement Reduces Maintenance requirements**
   - **c) Reduced Gut and Liver Size Reduces Maintenance requirements**
3. **Calf Weaning & Preconditioning made easy**
4. **Easier to apply all types of technology**
5. **Major disadvantages - If land acquisition, accumulation and possible appreciation is your goal, owning ranch land is a good excuse to own cows and visa versa.**

Cow Trading in a Drought (Better to buy than sell)

Buyer Advantages

1. **Cheaper cows because of depressed market, poor body condition and lighter weights**
2. **Poor condition bred cows may be low maintenance cows**

Buyer Disadvantages

1. **Shortage of grass – Dry lot alternative**
2. **You may buy someone else's problems (Bad temperament, health problems, etc.)**
3. **You may be the new kid on the block**

–K–

Future Expansion

1. **When Will The Cow Herd Expand?**

2. **Where will it expand? (Feed Follows the Water, Cattle Follow Feed)**

3. **Should the Cow Herd Expand? (Cattle are getting larger and beef demand may be declining)**

Many base herd expansion forecasts on weather, numbers and economics only. True, increasing expenses and the drought are important. However, expansion or lack thereof is also largely determined by other factors.

This includes an aging rancher population made up of individuals who have spent a lifetime in what was previously essentially a break even business with occasional profits coming from land appreciation. The old adage that "A Cow Man who dies rich has died before his time" contained more truth than fiction. Due to current higher land and cattle prices many may opt to "take the money and run".

A major factor which some may like to "sweep under the rug' is many cowmen are suspicious of current consolidation occurring in the packing and Feedlot sectors and what this could do to their bargaining power. Cowmen may not be the smartest guys in the room but, they didn't just fall off the turnip truck. They can see what has happened to the fat cattle market and the independent feedlot operator where often a "cash market" is non-existent. The old ten year cattle cycle was a disaster for many cowmen and they don't want to see it return.

In spite of significant profits cow numbers have steadily declined since the beginning of the millennium. This sends a powerful message that the cow-calf industry is undergoing significant changes.

–K–

Final Thoughts

1. Times are changing and always will
2. Be flexible and have an exit strategy
3. Listen to, but be skeptical of expert opinions

Expert opinions in late 70's (I was there) included:

a) Global cooling will be a major problem
b) World population will peak at 6 billion, then decline
c) No more grain fed to cattle

as has converted almost entirely to cow-calf production. Ron is a long-time friend and client with a keen analytical mind. Confinement cow operations have several advantages. After our drought wreck in Central Texas a few years ago, I moved most of my Texas and Oklahoma cows to Nebraska to Coe Cattle in Madison and Antelope Counties. Ed McClymont and Bill Emrich of Coe Cattle Company and I also partner on a number of Nebraska cows. This has worked well except for the extremely high feed costs in early 2013. Caroline and I actually began dry-lot cow production in the early 2000s because we often brought cows from drought areas and had no pasture available. We kept quiet about this program for several years because we didn't think many would be interested, and more important, we didn't want the competition buying cows. It obviously works better when feed prices are low, but it has advantages whenever you are short of pasture and hay. Among the advantages of confinement is reduced maintenance costs, improved calf health and much easier weaning. It shouldn't be considered a replacement for good pasture or range conditions, but in a drought you can often maintain cows better in dry-lot than you can feed them on pasture. As many have learned the hard way, you can't economically feed your way out of a bad drought buying expensive hay for "outside" cows. A major advantage of confinement cow production is you can increase your herd during a drought when cows are cheap.

The combination of low cow and total cattle numbers in the Southern Plains plus a substantial number of empty pens could be an opportunity for some feedlots. Confinement bred heifer and cow production would not only give us a needed increase in breeding stock, but it would also help fill empty pens. I've had several inquiries from feedlots, and some are building large confinement cowherds. The dairy industry has utilized confined heifer and cow programs for years. With lower feed costs and higher bred heifer and cow prices, this also has potential for the beef industry. This is a big departure from conventional feedyard operations, but those who are successful in this industry are those who can think outside the box.

In November 2013, I sold the last of my Nebraska farmland. Thanks to continued strong land prices that worked out well. Once again I'm out of the farming business which is probably for the best. In my world, equipment and machinery are the devil's tools. I'm better suited for the cow and timber business. Other than working out at the gym, I don't do much physical labor anymore. I've got some old injuries from my misspent youth that slow my movement so I tend to get in the way. My help in the corrals is like having two good men missing. Fortunately, we have a good crew that takes care of all phases of our business. With time on my hands, I tend to spend more time thinking "outside the box" and this has paid a few dividends. In retrospect, looking back at my early career, I'd have probably been ahead if I'd have spent less time working and more time thinking.

Because of phenomenal increase in land prices, grain prices and cow-calf prices the farmers and ranchers have been the beneficiary of the greatest boom these industries have ever experienced. Unfortunately, whenever there are big winners, there is often a big loser—and in this scenario it was the feedlot operator. Not only did feedlots have to pay more for their grain and replacement cattle, but many large feedlots didn't have a large land base, and therefore didn't benefit from the land boom. Furthermore, the large excess of feedlot pen space added to their misery. The recent rapid increase in fat cattle prices and decreased grain prices has helped heal the feedlots but they need more good times to recoup their losses. Lower grain prices mean farm income will drop, and it will be interesting to see the impact of this on farmland prices.

Beginning in the 4th quarter of 2013, feedlots again became profitable. Furthermore, the fat cattle market continued to improve throughout the 4th quarter, and ended a record high of $137 or $138 in late December. Unfortunately these feedlot profits were preceded by almost two years of substantial losses. The industry has a long way to go before their losses are recovered. The cow-calf and yearling sector of the industry also had a very profitable 4th quarter as did the dairies. For all segments of the cattle industries to be profitable simultaneously is indeed a rare occurrence and to be appreciated and savored. Calf, yearling, fat cattle, bred heifer and cow prices all reached record levels.

As good as the late 4th quarter was, it was only a warm up for early 2014 when price increases became breathtaking. In January 2014 fat cattle prices set new record highs each week with a top of over $150/cwt. Wholesale beef prices also reached records highs, and many are concerned about "demand destruction." Time will tell. Light calf and yearling prices also shot upward as people scrambled for pasture and feedlot replacement cattle. No doubt we'll reach a fat cattle top at some point and the consumers will tell us when. Competitive poultry and pork prices could also be a beef market challenge. However, PEDv continues to trouble the hog industry and it is estimated to be responsible for the loss of one million piglets.

It's interesting to contemplate the future for the cow-calf segment. It appears modest expansion is underway, and may continue. However, it may not be the traditional type of expansion as it may take place in confinement and semi-confinement facilities associated with feedlot and farming operations as opposed to pasture-based ranches. At these cow-calf and land price levels plus lower feed costs, intensive confined operations can be profitable and offer efficiencies which can't be realized in pasture operations. Furthermore, there is plenty of excess pen capacity available and I can tell from my phone calls many are considering this option. Finally, it is difficult to justify costly cows and ranches in droughts and environmentally stressed conditions. The current severe drought in the Southwest reminds us the drought is not over.

The long awaited cattle inventory report was released January 31, 2014, and pundits concluded "it was as expected with no surprises." However, I consider it bullish for cows, calves and yearlings. This is good news for ranchers but bad news for feedlots that can anticipate continued costly replacement cattle. Total cattle numbers were down 2% and at their lowest level since 1951. Total cow numbers were down 1%, and in sixteen of the past eighteen years beef cow numbers have declined. The combined total replacement cattle outside feedyards was down 3% indicating tight replacement supplies for the rest of the year. Even with the large decrease in cow kill in the 4th quarter of 2013, beef cow numbers still declined and the herd will be slow to rebuild. The strong packer cow and heifer market may bring more females to town resulting in even slower herd rebuilding. Packer cows over $1 per pound, and some believe the price will go substantially higher. This is an incentive to continue culling. The drought is not over in the Southwest. The past year is the driest ever recorded in California. Near the ranch I sold in the Central Coast area, rainfall the past twelve months has been approximately ½ inch which is barely enough to support the wildlife.

However, it's a mistake to blame all cow liquidation on drought conditions. For example, in the mid-South and Southeast where we had good moisture the past year in most states cow numbers decreased. Drought is part of the equation, but I think a larger factor is we have changed the culture of the cattle industry. Reasons include aging ranchers, alternative land use and some opted to take the money and run. Government programs have encouraged farmers to convert pasture land to crops and ranches have become farms. This may not be a desirable trend, but it's not the first time Government programs have had undesirable consequences. The trend for ranchers in the 35-55 age category to exit the industry is not a sign of a growth industry. Personally I'm not so sure we will increase cow numbers this year. Record high prices for heifers and packer cows is an incentive to liquidate. Furthermore, those fortunate to have a cow herd may not want to increase numbers at current prices, and those currently without cows may encounter financing problems when trying to build a herd. Finally, some older cow men may simply opt to sell and take the money and run.

What does this all mean? Most likely there will be continued good times for cow-calf operators and continued high break-evens for feedyards. We may see cowherd expansion, but not the traditional type. It will be in the higher rainfall areas with operations associated with feedlots and farming rather than the traditional ranches. Feed follows the water, and cattle follow the feed. Numbers might also increase in the Southeast as they harvest more CRP timber acres and replace it with grass.

Fortunately, I decided to extend this book, so this ends up as a 52-year as opposed to 50-year history, and best of all, it ends at a time of record prices in the industry. The current markets are the events of which dreams are made. Furthermore, we may be entering a new era, and I'm betting there's a young guy or gal just beginning their career who will someday write the next 50-year history. It is great to be a participant in our industry at a time when history is being made before our eyes.

With lower beef production come higher prices, and thus far the consumer response is somewhat surprising. Perhaps with lower beef production, demand becomes somewhat inelastic and prices are not the major factor for the consumer. The popularity of many high dollar branded beef products indicates many consumers are willing to pay a premium for high quality beef. If so, perhaps the industry can learn a lesson. In the past whenever we reduced production and became profitable, our reaction has been "let's expand." We often continue to expand until we again became unprofitable. In retrospect, it sounds foolish doesn't it? Again, the lesson may be "bigger is not necessarily better."

Domestic and export beef demand in 2013 and 2014 has been outstanding and indicates our beef check off dollars are paying dividends. Worldwide protein and beef demand is exceptionally strong and the meat export federation has done a good job of taking advantage of this. Randy Blach and Cattle Fax group have had a hot hand in the market projections the past several months. I asked Randy what he thought the major changes in the beef industry would be in the future. With hesitation, his answer was the increased export demand for beef, meat and protein. To illustrate this he pointed out that in 2013 we exported approximately 16 billion pounds of meat and we are on track to increase this substantially in 2014 and it could reach 20 billion pounds by 2020. On a per head basis, export demand has added $308.00 in 2013 and we're on track to go to $355.00 per head in 2014. Compare this to $175.00 per head in 2003. This growth trend is nothing short of spectacular. Growing world population and affluence should bode well for the future.

If Randy is correct and the prime mover of our future beef demand is export business, this presents a bit of a challenge for some in our industry. Some ranchers are inclined to live in their own world and it's sometimes difficult for them to consider what's happening out of state or even out of the county. Asking these people to factor in events in China, Indonesia, India, etc. requires a significant change. This doesn't suggest they're ignorant or backward but, they are accustomed to surviving in their own unique local environment. That may be a county, state or even a multi-state region but you don't hear much conversation at the local coffee shop concerning beef demand in China, India, Katmandu, etc. Future coffee shop talk may contain comments such as "I see they're going to feature ground beef in Beijing, China next month. Maybe we should preg early and send a few more cull cows to market."

One of my most requested poems is "Good News Overdose" which reminds us when things look best, unexpected events often

Good News Overdose

K. S. Eng

The harvest was good & feed prices low
The pastures are green & the creeks all flow
Health is no problem cause the pens are dry
Performance is great & cattle prices high

The bankers relaxed & sit there smiling
I feel nervous; cause there's no reason for lying
He grins and asks, "Can we lend you more money"
I should be happy, but I feel sort of funny

There just something missing, a mysterious void
I've called my physic & I'm reading on Freud
I'll write Ann & Abby when I get back
If that doesn't work, I'll call Baxter Black

I talked to my bartender who's experienced & loyal
We had several sessions, along with Crown Royal
He said "you're like a cow given too much hot feed
You've had too much good news & now
you've O.D"

Your accustom to bad news, it may sound strange
But, the business you're in, good news will change
I drove back to the ranch, met the wife at the gate
She said "we've got problems & you're 7 days late"

"The truck is stuck in the furtherest field
The buyer is mad cause the cattle didn't yield
The banker called & they hired a new staff
They took your values & cut them in half"

"Our foreman is in the hospital, got thrown
off his horse
The cook went home and filed for divorce
The vet looked at the sick calves, the
advice he gave
It's too late to worry cause there's not
many to save"

She said, "let's go to bed, you've done
enough harm
The furnace isn't working & you can
keep me warm"
Then she whispered in my ear about
some IRS letter
It must be the bad news; I'm starting to
feel better

blindside us. Examples of this would include the beef price freeze in the early '70s, the dairy buyout in the late 80s, BSE in 2003 and Pink Slime in 2012. Today we may be in the midst of another good news overdose, but what a ride it has been. It's a good time to try to be humble, But as the Mac Davis song says, "Lord it's hard to be humble." Risk and projected breakevens have never been higher, and if it's hard to be humble, at least consider that you may not be bulletproof.

When I extended the book to early 2014, I hoped things might turn around by then. Did they ever! I'm an optimistic person, but I never dreamed of these profits for almost everyone in the cattle industry. It proves once again that nothing beats good timing and good luck.

If the cattle and land business is your game, it has never been better. Cherish the moment, hug those close, and count your blessings.

I'd Do It All Again

K. S. Eng

I dreamed I stood before the judge
He asked "Ken how do you plead?
It appears that you've been reckless
And sowed some wild seed"

I said, "Once I was a little rowdy
And I've drank my share of wine
There were times I've acted stupid
Would you like to see my sign?"

"Cattle have been my business
And I've had a wreck or two
But if I have any regrets
It's about things I didn't do

"I've had a fondness for the ladies
Sometimes I was careless with a heart
But some treated me the same
And we were friends when we'd part"

"I was crazy in love with travel
No matter what the cost
"I'd go to the middle of nowhere
Because it felt so good to be lost"

"I wasn't good at long range planning
And though my plans lacked
perfection
When life didn't go as planned
I could adjust and change direction"

It's been said, "you have a temper
Is that true asked the Judge
I said, "There're times I get
aggravated
But I never hold a grudge"

"It's true I didn't sweat the small stuff
I hope that doesn't make me a fool
I've ignored petty paper work
But I obeyed the golden rule."

It says here your clothes
seldom matched
The Judge asked, "How do you plea?"
"Judge," I said, "I can't deny it
I plead guilty in the 1st degree."

The Judge said, "You're obviously
not perfect
Now that you've heard the
evidence, Ken,
What if I gave you a second chance?"
I replied, "I guess I'd do it all again"

-EPILOGUE-

Looking back, it has been an incredibly exciting journey. I have been blessed to travel with lady luck, good friends, good clients and most of all Caroline.

When I come to the end of the trail, plant me next to Caroline at Shiloh Cemetery, toast us with a good glass of wine, and recall the stories you can tell.

Acknowledgements

As always, Annie Powell rendered invaluable assistance in many ways to make this project happen.

Citation to the Editor

As I was wrapping up my book, Baxter Black advised it was extremely important I get a good editor. He said I know you think you're a good writer because lots of people have told you so but, don't even attempt to publish the book without an editor. A good editor will not change what you say but, enable you to say it better.

The original lady I had in mind came highly recommended but, was committed to other projects when I contacted her. However, I got lucky and Terry Sherrell fell into my lap (figuratively, not literally). She is with our production company, and I asked her if she had someone she might recommend as an editor. Her reply was she might like to try herself. I replied, why not.

The results were magic and she made changes and suggestions that improved my original manuscript and never once damaged my pride. As Baxter suggested, she didn't change what I said but, she said it better.

Terry, thanks for all your help and once again, I just got lucky.